WISDOM *for* BEEKEEPERS

WISDOM *for* BEEKEEPERS

500 TIPS FOR KEEPING BEES

JAMES E. TEW

BLOOMSBURY

LONDON · NEW DELHI · NEW YORK · SYDNEY

Published 2014 by Bloomsbury Publishing Plc
50 Bedford Square, London, WC1B 3DP

www.bloomsbury.com

Copyright © 2014 Quid Publishing

Conceived, designed and produced by
Quid Publishing
Level 4, Sheridan House,
114 Western Road
Hove BN3 1DD
www.quidpublishing.com

Designed by Clare Barber

ISBN 978-1-4729-0003-6

Every effort has been made to ensure that all of the information
in this book is correct at the time of publication.

Printed in China by 1010 Printing International Limited

1 3 5 7 9 10 8 6 4 2

To see our full range of books visit www.bloomsbury.com

To see our full range of books visit
www.bloomsbury.com

To beekeeper friends,
past and present

CONTENTS

INTRODUCTION

Keeping bees is a mixture of craftsmanship, artistry and luck. For all our years of study and caretaking, beekeepers can still only comprehend a small part of the bee's mysterious world. Every day, beekeepers and scientists fill in some pieces of the unfinished honeybee puzzle, but it's far from complete. The idea of this book is to provide a reference source for the novice beekeeper. The 500 practical, down-to-earth tips will guide readers through the basics of caring for their bees. We are devoted to our bees because of the sweet honey they provide, but most of all, we are drawn to the joys that keeping and tending to them can bring.

BECOMING
A BEEKEEPER

While anyone can become a beekeeper, the important thing to establish fairly early on is if the hobby is right for you. The beekeeping craft can be practised from a small domestic level up to a large commercial operation. For those who keep bees, the craft is challenging and diverse enough to provide growth and enjoyment for a lifetime.

FIRST THINGS FIRST

TIP 1: *Brush up on your bees and honey*

🐝 Long ago, honey drew humans to bees and this natural sweet food was highly coveted: it was either eaten on the spot or honeycombs were taken back to the family group. Protective gear did not exist for early bee-loving humans and bee-smoking devices were usually nothing more than a smouldering torch. Honey hunters clearly had to be tough enough to take stings.

🐝 Honeybee brood (see tips 497–500) was a high-value food source and was eaten along with the honey.

🐝 Wax has always been used for high-quality candles and as an early waterproofing agent.

🐝 The oldest known alcoholic beverage is fermented honey (mead). The demand for this product must have made culturing bees even more lucrative.

🐝 Early observant farmers knew that bees were important for pollinating plants. That appreciation has grown to the extent that honeybees are the primary pollinators of many food and fibre crops today.

Though all bee species are vitally important, honeybees have become established and are nurtured around the world. They have always been invaluable for providing and producing food. Beekeepers and the bee colonies they manage provide an essential service to society.

TIP 2: *Establish whether you are the beekeeping type*

🐝 For those who are a 'good fit' for beekeeping, the craft is a comprehensive, fulfilling experience that can provide lifelong enjoyment. Some of the interests and concerns among good beekeepers are: an appreciation of nature, biological curiosity, interest in gardening or farming, concern for a healthy ecosystem and a fundamental interest in helping others and in helping bees.

Though a small industry, beekeeping is remarkably diversified and is not just about honey production. Pollination services – both small and large – are always in demand. Bees are excellent biological models for teaching students biological tenets at all educational levels. Many diverse types of honey can be produced. On a more practical side, woodworking is a fundamental component of beekeeping; metalworking and plumbing skills are important for setting up honey processing equipment too. Beekeeping complements gardening – both flower and vegetable gardens. Cooking with honey is a popular culinary interest.

To a degree, anyone can pursue an interest in beekeeping or in aspects related to beekeeping. Not everyone has to actually start a colony. Beekeeping or related beekeeping activities can be enjoyed in many ways.

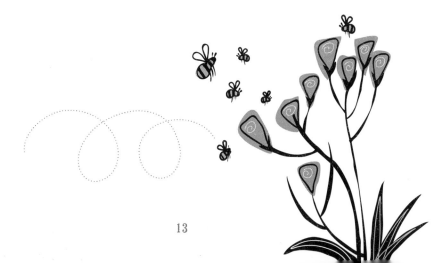

TIP 3: *Remember, there is no 'standard' procedure*

🐝 At the outset, like any unfamiliar endeavour, beekeeping may seem confusing. A person who is curious about beekeeping and wants to know a bit more should start with basic reading. Bee information is everywhere – the web, the library or magazines. First, get the gist of the craft and its terminology (see tips 17–25). Beekeeping clubs are a wonderful source of information, so check for the closest to you. Contact the British Beekeeping Association for the names of some local beekeepers. Talking with those already in the hobby will help enormously. Early on, expect different opinions and recommendations. There is no actual standard beekeeping procedure that covers all situations. Feel your way along and make the most of local resources. It will take the average person about a year to become comfortable with the basic concepts of beekeeping. (See page 288 for a list of websites and groups to help you get started on your beekeeping journey.)

TIP 4: *...But beekeeping is built upon broad principles*

🐝 Though a bit confusing at the beginning of your endeavours, the broad principles of beekeeping will quickly become obvious to the new beekeeper. Some of these principles are: the division of labour within the colony (see chapter 4), the design and function of the standard hive (see tips 33–44), the typical activities of bees during the four seasons (see chapter 8) and how to manage the occasional stinging experience (see tips 162–166). An understanding of broad principles like these and an acceptance of the many questions that still remain unanswered will pave the way for you to develop into an accomplished beekeeper.

TIP 5: *Try to choose a level you're comfortable with*

🐝 Becoming a beekeeper is not a specific event or a fixed thing. The number of colonies kept and the type of beekeeping that is undertaken have a bearing on how much time and work will need to be invested. Keeping a few hives – maybe three or four – of bees is not particularly physically taxing, but clearly adding more colonies will require increased labour from the beekeeper. Declining health or reduced general fitness generally doesn't mean that a beekeeper needs to give up the craft, but instead they may decide to carry on in a different direction. For example, rather than producing large honey crops, the experienced beekeeper may seek out other rewarding activities such as deciding to teach a beekeeping class or to experiment with queen production. The list of related interests is endless.

TIP 6: *Be prepared for occasional hard work*

New beekeepers should grow their colony numbers fairly cautiously to allow all facets of the budding operation to grow at a manageable rate. A few colonies are enjoyable, but more than 10–15 colonies can become quite a chore. The basic issue with the standard beehive is that the height may change depending on the season. During winter months it's it may just consist of two brood boxes, so will be low to the ground. If that colony is to be inspected, crouching beside it or stooping over it becomes necessary. If it is raised up sufficiently to remedy this situation, it will then be too high to add and remove 'supers': Supers are boxes with frames of comb used for storing surplus honey. Basically, be prepared for some stooping and twisting.

TIP 7: *You'll need to wear protective clothing*

Protective clothing can become heavily soiled with wax, propolis (a sticky resinous mixture that bees collect from things such as tree buds and sap: see tips 463–475), honey and general grime. The dirty suit is simply doing its job. Use a commercial washing machine to launder heavy suits rather than using a home washing machine . The wax and propolis will not readily wash out, so the protective clothing will quickly develop some 'character'. The suit protects the beekeeper from some of the bee product stains and smoke odour. This is not a major issue, but there can be no ignoring the smoke odour that the beekeeper's clothes acquire.

TIP 8: Expect some stings along the way

If one keeps bees long enough and even wears heavy protective clothing, sooner or later a sting or two will occur. A requirement for becoming a beekeeper is the ability to tolerate the occasional sting. Generally, the beekeeper who works colonies several times per month during warm months is stung a few times on each trip and will develop a tolerance for stings. Initially, some local pain and swelling will probably occur, and it may take a day or so for the symptoms to fade. Some stings are worse than others. The amount of venom administered per sting depends on the age of the bee and whether or not she has used her sting before. Even if many hives are being manipulated, significant stinging events are rare. If the bees are particularly defensive, review the conditions and determine what the causes are. Simply applying more smoke may not be the best solution (see tips 162–166). Life-threatening sting reactions are rare, but if anything looks or feels wrong about a sting or stinging event, contact your doctor. A good rule of thumb is to watch for symptoms that occur away from the site of the sting. For example, a sting on the hand should not result in swollen eyes or a rash on another part of the body.

TIP 9: *Don't be deterred by endless allergy stories*

Stinging behaviour, no doubt, is the primary reason why most people do not have an interest in bees. So there it is…stings are not enjoyable. But trained and experienced beekeepers have the protective gear and knowledge to deal with the occasional sting. The normal sting will result in some redness, local pain and swelling. While that is a reaction, it is not considered to be a harmful reaction. When others hear that you're a beekeeper, expect to hear an alarming sting allergy story from them. While beekeepers know better, most beekeepers are not trained doctors. Even though you may be sceptical of hearsay, if a non-beekeeper asserts that they have a serious reaction to a sting, take them at their word. Accept the story and encourage the person to see their local doctor. The rare, but authentic, allergic sting reaction is frightening and serious.

TIP 10: *Realise that support from family and friends will vary*

To most people who do not keep bees, apiculture is respected, but just a bit unusual. Acceptance by family and friends varies greatly. In the husband/wife relationship, normally only one chooses to become the keeper while the other, by default, must become the supporter. Just as in society at large, extended family support (or lack of support) will vary widely. Understand that this new interest is something that you are pursuing and do not expect them to necessarily follow. However, sharing your honey crop (see chapter 9) may attract their interest!

TIP 11: Embrace the social 'Bee Fever' benefits

🐝 For those who have developed an interest in beekeeping, the passion shown for the craft can become overwhelming for those around them who aren't interested in beekeeping. Beekeeping eagerness can be hard to control and normal conversation can become one-dimensional. At first, family and friends may find it funny and entertaining, but that phase passes. 'Bee Fever' would describe the state you're in at this stage. Ask friends and family to be patient with you. Even though this compulsive passion subsides after a few months, beekeeping continues to be a source of conversation when talking to others. 'So, you're the bee man (or bee woman)?' is a frequent conversation introduction at a social event.

TIP 12: Introduce children to the craft with care

🐝 Once an adult has a beekeeping interest, children in the family will frequently become curious about the project. The beekeeper must show good judgment before exposing children to bee colonies. Just as with adults, the personalities of children will vary widely and there are no specifics on how old a child should be. Protective clothing is available in children's sizes, just ask your supplier.

Start very slowly in short bursts. At first, maybe let the child puff a new smoker or try on a veil. If possible, dress the child in protective attire and go for a quick walk in the apiary without actually opening a colony. If available, you could show them a laying worker colony – a colony that is without a queen and therefore small and fairly harmless – or a small mating swarm that was recently hived.

TIP 13: *Be mindful of neighbour concerns*

❀ Neighbour conflicts can be challenging at times, so it is best to start a dialogue early and prepare your neighbours for the arrival of bees. In defence of the non-beekeeper, the beekeeper's bees can be something of a pest. For example, it is impossible to keep bees from exploring the neighbour's birdbath and, occasionally, a swarm will escape to the neighbour's property; some beekeepers also have issues with their bees defecating on buildings and cars. While it is true that suburban honeybees provide valued pollination services, there are a few less desirable attributes with which neighbours must deal. Respect your neighbours at all times.

TIP 14: *Stay clear of property boundaries*

❀ As much as possible, keep your hives away from boundaries. This may be a bigger problem in urban settings than in rural areas. High fences or barriers formed by trees and shrubs will force your bees to gain altitude quickly. Painting your hives a colour other than white can be helpful, as will keeping them in the backyard, not out in the front garden. A hive that isn't obvious won't be as noticeable to the people living next door.

TIP 15: *Educate others about honeybees versus wasps*

Your bees will be 'based' in your apiary, but they will travel over several thousand hectares that surround your property. This extensive foraging behaviour is a good reason to stay away from other people's boundaries and to prepare neighbours for the swarm that may issue in spring. A greater issue may be other bees or wasp species that have crossed the channel recently. Wasps and carpenter bees are notorious for causing honeybee-keepers problems. (Wasps normally live in underground nests while carpenter bees bore holes in trees and fascia board.) Neighbours several houses down may blame your foraging bees for stinging them when it is yellow jackets that are issuing from a ground nest. Though beekeepers know there is a big difference in species, neighbours may feel 'your' bees are burrowing into the roof space. Be patient with others and, when possible, educate them about the differences.

TIP 16: *Check out the regulations*

Regulations need to be followed or you and your bees will be caught short when someone complains, though the attitudes of town or city councils vary extensively (see page 288 for associations that may help you find out more). Normally no harm is done by contacting officials directly, but you may first want to ask other established beekeepers what the rules are for keeping bees in your area. When officials are contacted, if no clear regulation exists, some confusion may result and new regulations may be initiated.

At times, an enthusiastic beekeeper pushes the regulatory envelope too far: possibly too many hives are kept on a particular lot or maybe children play near the apiary. Essentially, don't create a problem where there isn't one, but always respect the 'spirit' of any existing regulations.

GETTING STARTED

TIP 17: *Judge how much time and dedication is needed*

✹ Time spent working a bee colony will vary with the seasons. The following estimates are offered. Working a single colony will take about an hour or two per week during spring and maybe an hour every other week during summer. As autumn approaches, time needed may creep back up to an hour or two per week, but during winter not much will be required of the beekeeper. Each subsequent colony added to the workload will probably add about another 30–45 minutes per colony. Different beekeepers will work at different speeds and have different standards, but in general, working one or two colonies does not take a significant amount of time. Allocate time wisely.

TIP 18: *... And how much money*

✹ There will be a cost for first-time purchases of veils, smokers, hive tools and protective clothing (see chapter 2 for more information on equipment) – around a couple hundred pounds. Hive equipment for a single hive will cost about £100–200 depending on what kind and how many are purchased. Obviously, unassembled equipment will cost less. The live bees will cost anything from £50 to £150. The estimated cost for starting a single hive in new equipment will be something like £200–300. Additional hives will require less money, since the smoker and related protective equipment will not be necessary. There are many variables such as postage costs, type and amount of equipment and whether or not it is assembled. Happily, bee equipment will last for many years if properly cared for.

TIP 19: *Become familiar with beekeeping vocabulary*

As with many other callings, beekeeping has many unique words and phrases. Hive equipment may have different names depending on how and where it is used. There will be a learning curve at the beginning, but the terminology will come quickly. All beekeepers were new to the craft at one time, so all experienced beekeepers will understand your situation. Don't be afraid to ask questions, and quiz others until you gather the information you need. With time, the terminology will become familiar to you.

TIP 20: *Source your initial information from books*

Beekeeping is an old, entrenched craft and books containing various opinions abound. The mechanics of the standard hive have not changed much, so an older book would still be relevant. However, bee pests and diseases have changed greatly, making an older book less useful in this area. Ask your new beekeeping friends what books they have found useful; ask around at your beekeeping club meetings; borrow books from the library, and use websites and forums that have been recommended.

TIP 21: *The Internet is useful, though limited*

The web is filled with beekeeping information. Unlike books, which undergo editing and review and require the investment of time and publication money, webpages can be turned out fairly quickly. On one hand, fast, readily available information is a good thing, but poor or biased advice will not be helpful to the novice beekeeper. At the beginning, explore pages posted by various universities or by the Department for Environment, Food and Rural Affairs (DEFRA), beekeeping suppliers, clubs and associations (see page 288). Later, as your experience grows, branch into exploring individual webpages.

TIP 22: Test out web-delivered classes and training programmes

In years past, various correspondence courses were available in printed form or photographic slide sets. Now, video presentations on the web show ways of doing nearly anything related to beekeeping. As with finding any web-based information, some of these videos are individually produced and may be of questionable quality, whereas others are great. While the Internet is a great medium and can offer excellent teaching systems, the new beekeeper will miss the true sense of being in a live beehive with buzzing bees all around.

TIP 23: Seek help from experienced beekeepers

Though there are books, webpages and DVD for teaching the fundamentals of beekeeping, without a doubt the best way to learn beekeeping is to watch an experienced beekeeper work a colony. There is no standard way to find a mentor. Even if you do find one, some are better than others. Beekeeping associations can be very helpful in this regard. Another alternative is to attend workshops where there are apiary demonstrations. Clubs, beekeeping suppliers and agriculture departments all provide them. These outdoor events are very useful to the new beekeeper. The main thing that comes from such demonstrations is how to smoke colonies and how to handle frames with confidence.

TIP 24: Select the site of your first apiary wisely

An apiary site, in many instances, is a temporary thing. However, other apiaries may be in place for years. In selecting your first site, get as far from human traffic as possible. Choose sites with the following attributes: availability of food sources, nearness to water and absence of pesticides. Barriers around the apiary are helpful – even necessary. There are recommendations in many bee books about facing the hive towards the southeast (for the morning sun) and to avoid the winter winds.

More often than not, the new beekeeper puts the apiary in the best spot available to them. While the site is certainly important, it is not the only important aspect of the new bee operation. The first apiary is a valued memory to the individual who becomes a lifelong beekeeper. Even if it wasn't an overly impressive apiary, it was your first.

TIP 25: Once underway, remember to pace yourself

For many new beekeepers, the craft is so enjoyable, they want to expand – more apiaries, more honey. Each beekeeper has a tipping point. The beekeeper keeping bees as an enjoyable hobby changes into the beekeeper keeping bees as an income, and the hobby turns into an obligation. Money has been invested and a trading identity established, and all too often the overloaded beekeeper begins to burn out, stops going to club meetings, and neglects the hives. You should also be aware that there are regulations, administered by state food authorities, which must be followed when selling any food stuff. At a minimum, registration is required. Larger bee operations aren't for everyone.

APPLYING A PRACTICAL APPROACH

TIP 26: Start with two colonies

For the new beekeeper, who is starting a project from nothing, two hives are manageable and generally provide plenty of honey. Since beekeeping is a biologically driven experience, some things can simply go wrong. Sometimes the new beekeeper is at fault; at other times, it's just bad luck. Having two colonies gives a backup colony to assist the ailing colony. Certainly, a single colony can be the starting number – especially if the new beekeeper has established beekeeper friends.

TIP 27: Stick with traditional hive equipment at first

Selection and implementation of beehive equipment is a fundamental aspect of keeping bees. Commercial bee boxes are nothing more than artificial bee housing where beekeepers entice wild bees to live. Standard equipment of standard dimensions meets the needs of most new keepers.

There are other styles of equipment that offer rewarding but different beekeeping experiences. Some examples of other readily available equipment types are: top bar hives and Warre hives (see chapter 2 for more information on hive equipment). Certainly, there is no reason not to try these alternative styles of equipment. Experimenting later in your beekeeping career will aid your success with this type of equipment. Unless there is a support group to guide you, the novice should start with standard equipment.

TIP 28: Be aware that beehives require some assembly

🐝 For the novice, the first-time selection of equipment can be confusing. Beginner kits are available from most beekeeping suppliers, and this is an appropriate way to go for a new beekeeper who doesn't yet have a beekeeper support group. The kit smoker may be a smaller one and the veil may not be the most deluxe model, but at least the new beekeeper gets all the basic component parts without having to select components individually.

Some suppliers supply fully assembled hives, although not necessarily painted. All the new beekeeper must do is open the packaging and install live bees. Although this may be helpful for getting the new keeper started, the cost of both the equipment and the packaging will be more expensive. Experienced beekeepers probably would not consider such an option, though commercial operations might find the service worthwhile.

TAKING IT TO THE NEXT LEVEL

TIP 29: *Invest during the first season; it may be honey-free*

Most new beekeepers will have a keen desire for a honey crop from their new hives. In most cases, this will probably not happen. If a colony is initiated from scratch and combs must be built, the new colony must invest significant food resources in setting up and getting ready for winter. Those resources represent the honey crop from the first season. In areas with shorter, more intensive nectar flows, honey crops are more commonly made from new hive units. Even so, most beekeepers in most areas should not expect a honey crop from a new hive during the first season. Remember that the new hive is still establishing itself and that, for most new colonies, the first winter will be a significant challenge. If the autumn nectar flow is not a good one, any honey taken from the spring crop may be a critical loss. Autumn honey is typically left on the hive for winter.

TIP 30: *Consider becoming an office-bearer in your club*

As your beekeeping skills grow and network expands, you may find that you're keen to take on one of the office bearing roles in your association. Beekeeping clubs, at all levels, are essential to the training and nurturing of new beekeepers, and a strong supporting club committee is always needed. You don't need to have expert beekeeping skills, as much as you need management and leadership skills. The workload will vary depending on club size and the number of events that the club runs. It can be hard work at times but beekeeping club officials always have the thanks and respect of the members.

TIP 31: *Go commercial, if you like*

It is not uncommon for an occasional beekeeper to grow the bee operation to the point that it begins to generate income. Much of this impetus depends on the specific conditions at hand. Local honey and pollination services if required in your area. Unless born into a beekeeping family, most beekeeping businesses are built at a steady pace for several years. As with other young businesses, many aspects of the growing business must be kept in balance (see tips 418–425).

TIP 32: *If you do want to grow, think of the labour required*

In the early years of a growing beekeeping project, much of the actual labour will be done by you, the beekeeper, with only the occasional help of some friends or family members. A few hives are enjoyable, but much more than 20–30 and real work evolves. As the years pass, this workload will increasingly become less practical. In some instances, paid labour will be needed, but the cost will probably require the operation to grow even more. Most small commercial beekeepers seem to reach equilibrium. Hive growth tops out and facilities and processing equipment evolve to the point where any further growth will need to be particularly profitable. These small operations can be rewarding. The job is performed locally and the owner becomes reasonably well known. For some, it is a good lifestyle.

BEEKEEPING EQUIPMENT

Beekeeping equipment is specialised. Specific pieces have acquired novel names that at the time were logical but now appear odd: supers, inner covers, deep hive bodies and reversible bottom boards are prime examples. While it may seem confusing at the outset, such terminology and the use of each piece of equipment will quickly become familiar.

BEEHIVES

TIP 33: *Understand various beehive designs*

New beekeepers would be advised to use the standard hive of their country. All standard bee hives contain frames of beeswax combs in each super. This hive design is the most common in use, despite dedicated supporters of variations such as eight-frame equipment (two fewer beeswax comb frames) or top-bar hives (where the frames are laid out horizontally across the top of the hive).

The image on the page opposite shows the common components of a 'Langstroth Hive'. This is the standard hive of the USA and many other countries. It is becoming an increasingly popular point of intrigue and many UK universities buy in this model. However, in the UK the most popular common hive is 'The British National Hive' (see tip 34 for more information on hive types specific to the UK). Depending on the season and the colony's space requirements, the hive may be taller or shorter at different times of the year (see chapter 8). Generally, the bees nest is in the deep hive body, while surplus honey is stored in the upper supers. The queen excluder confines the queen to the lower deep hive body so brood and honey will not be mixed. The deep hive body is often called a 'brood box'.

Major components of a honey beehive

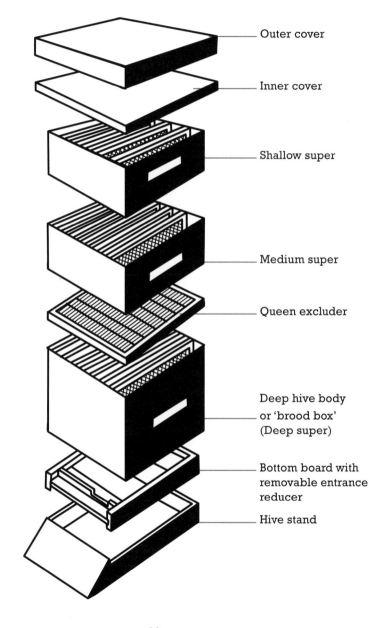

Outer cover

Inner cover

Shallow super

Medium super

Queen excluder

Deep hive body
or 'brood box'
(Deep super)

Bottom board with
removable entrance
reducer

Hive stand

TIP 34: *Know the available hive options to your country*

In the UK, beekeepers have many choices of beehives. The 'Modified Commercial', 'William Broughton Carr (WBC)' and the 'Smith' are just a few. However, the 'British National Hive', also known as the 'British Modified National', is the most preferred hive and used the most throughout the UK.

The National suits non-prolific or native type bees, and is said to be well suited to the climate in the UK. The frames are virtually interchangeable to the WBC, Commercial and the Smith apart from the shorter lugs. The frames are known as British Standard (BS), with the brood chamber and the super holding 11 British Standard frames. These frames can be normal frames or changeable to the 'self spacing' frames known as Hoffman frames. There are three standard depths of frame known as shallow holding 140mm (5½in) for supers, or 'shallow frames'. There are two types of frames for the brood chamber – the brood chamber is 215mm (8½in) for standard depth brood boxes, known as 'brood frames'. Then there are larger frames at 304mm (12in) for deep brood boxes, known as '14 x 12', or 'B.S. deeps'. These are sometimes erroneously referred to as 'deeps'. Some beekeepers prefer these deep frames especially if they have a prolific queen. The roof can either be pitched or flat. Pitched is used for its grandeur but flat roofs are the most popular.

TIP 35: *Remember, no hive design is perfect*

Every hive design has some flaws. Overall colony weight, when a hive is full, is significant and will generally require two beekeepers to move it. Additionally, the weight of individual components of a hive can be remarkably heavy, requiring individual frames to be moved rather than the entire heavy box. For instance, a deep super, when full, will approach 45kg (100lb). All that weight is lifted by two 1cm (⅜in) deep handholds using only sticky, glove-covered fingertips. Some beekeepers nail narrow wood strips on the hive ends to serve as a larger handle. Oddly, no standard attachment has been developed for attaching the outer cover to the hive top. Therefore, a brand new beehive will need to have a rock or a brick put on top to keep it in place.

TIP 36: *Use shallow reversible bottom boards*

One common style of bottom board is reversible – 2cm (¾in) on one side and 1cm (⅜in) on the other side. Normally the deep side is used during the summer and the shallow side during the winter. Use the shallower entrance year-round. The bees don't mind, and it saves you the extra effort.

TIP 37: Select hive designs with reinforced corner joints

Full bee boxes taken from hives and loaded into a truck are put under stress when full. It is easy for the corner joints to flex, resulting in the bee super becoming loose. Most commercial beehive manufacturing companies build bee boxes that use box joints at the corners. These joints improve the structural integrity and make cross-nailing easier. However, these joints can be challenging to cut in the home shop. A commercial or homemade cutting jig will be required to safely and accurately cut these joints. However, if the hives are rarely moved and only a few boxes are being used, it is common for simple butt joints or joints to be used for hive construction. In reality, a bee box only lasts for about seven years and is not a piece of fine furniture. Even so, box joints are preferable.

TIP 38: Don't be afraid to try plastic hive components

The international beekeeping industry has been tinkering with various styles of plastic beehive and hive components for many years. Most designs of early plastic hives were not suitable, but in the past decade improved plastic hive and hive components have become more accepted by beekeepers. Initially, bees don't care for plastic combs and foundation inserts, but once the first cycle or so has passed, used plastic equipment is readily accepted. The advantage with plastic frames is that they are complete. Simply open the shipping container and insert into the bee box.

TIP 39: *Use wood glue and nails to assemble wooden frames*

✿ If you still prefer wooden frames – and many beekeepers do – frame stability is greatly improved when common water-resistant glue is used. In fact, the nails are primarily useful in holding wooden parts together while the glue sets. Though a bit luxurious, if many wooden frames are to be assembled, a pneumatic pin nailer that drives 3cm (1¼ in) 18-gauge pins is really useful. A single nail driven through the end bar into the thicker part of the top bar combined with wood glue makes the joint nearly indestructible.

TIP 40: *Paint the inside of the hive to protect the outside finish*

✿ Just as nails for assembling equipment and plastic hive parts have evolved, so have the quality and selection of paints that are suitable for hive coatings. In many old beekeeping texts, there is an admonishment not to paint the inside of the beehive, but, with new latex finishes, that is no longer a major concern. If hive insides are not painted, moisture will wick through the box walls and cause the paint film to fail from the inside – not the outside – of the hive box. Be sure to let the paint cure, not just dry, before using it on the hives. You should know that, historically, inside surfaces of most hive bodies have not been painted; therefore, painting these is not a hard requirement, but does provide a more durable, protective paint film.

TIP 41: *If you prefer, use only one size hive box*

✿ To simplify beehive equipment, it is not uncommon for beekeepers to use only one size box. Normally, deep hive bodies are used (which are heavy) or deep shallows (17cm / 6⅝ in deep). In this way, frames and boxes, and hive loads, are standardised. The bees don't seem to have a notable reaction to the size selected. The shallower box (17cm / 6⅝ in) is desired to lighten the weight that the beekeeper carries when removing honey.

TIP 42: *Know what a queen excluder is, and use it*

In the bee-management world, queen excluders have historically had mixed reviews. Simple in concept and design, they are intended to be devices that limit access to workers and to a specified hive area. The open space within the grid should be in the range of 4.22mm–4.3mm ($^{11}/_{64}$ in–$^{3}/_{16}$ in). The intended use of the queen excluder is primarily to keep the queen bee from putting brood in honey storage supers (see tips 145–151 for more on queen bees, their role and information about other bees in a colony). The principal argument against using an excluder is that large, healthy honeybees with a full crop of nectar will have a difficult time squeezing through the excluders. If the spaces in excluders are made just a tiny bit larger to allow loaded workers easier passage, slightly undersized queens can potentially squeeze through these spaces. Since drone's thoraxes are even larger than queens, emerging drones above an excluder are trapped there.

Always looking for an improved design, inventors have presented many styles of these devices. None have proven perfect. You can occasionally use wood-bound welded-wire grids as opposed to punched sheets of plastic or zinc. There may be problems with drones and potentially restricted honey crops, but you will not have to contend with brood in honey supers.

TIP 43: *Provide a solid footing underneath your hive*

There has never been a universally accepted hive stand. Bee supply companies have sold great numbers of simple wood stands with a slanted front. This places the hive near the ground and provides a good place for vermin and pests to hide beneath the hive. Additionally, since these stands are in direct contact with the ground, wood rot begins within a few years. Probably the most common hive foundation is two common cement building blocks, which work reasonably well. The problem is they are just a fraction of an inch too narrow, the beehive just barely sits on these blocks. At first, this is a fine arrangement, but as the colony grows and puts on weight – lots of weight – and as rain comes to make the ground soggy, the hive can begin to lean. Even if the stacked hive does not fall over, it can tilt backward, allowing rainwater to accumulate within the hive. One solution is the use of two treated 10cm (4in) posts that are 1.8m (6ft) long that sit on two half-blocks. You can put two hives on this arrangement. There is nothing special about this idea and many other designs exist. Observe what other beekeepers use to create a solid foundation.

TIP 44: *Go for heat-branding wooden hives, if you can*

Since all beekeepers use pretty much the same style of equipment, unless there are clear identifying marks, ownership can be unclear. Wood branding the equipment with either a name or the state registry number is a wise move. For most beekeepers with just a few hives to mark, buying this device is probably impractical. A beekeeping association can buy the branding iron and gas bottle, but still the individual beekeeper needs to order their own branding head to attach to the blowtorch. The good news is that if one is purchased, it will last nearly forever. Some prototypes for GPS marking devices allowing beekeepers to mark their colonies in nearly invisible fashion have been researched and will hopefully soon become available.

PROTECTIVE CLOTHING

TIP 45: *It's best to start with a heavy protective suit*

❀ The evolving beekeeper will probably need two to three styles of protective clothing. There are many from which to choose. For the new beekeeper, the heavy-duty full suit with the zip-on veil is the best first choice. It is a suit that will withstand all bee events (and it is the very same type of suit that experienced commercial beekeepers choose, too). Leg and arm openings will need to be closed. Hook-and-loop fasteners (Velcro®) or simple duct tape are commonly used to secure trouser cuffs and wrist openings. Extensively reinforced suits will prevent most stings.

TIP 46: *Be prepared for the heat of hot summer days*

❀ As a beekeeper, heat is more of a problem than stings. The heavier and more secure the bee suit, the hotter heavy clothing will be on warm days. Dressing lightly underneath can result in perspiration soaking the heavy suit, making it a bit easier for bee stings to penetrate it. Different methods of ventilation and cooling devices have been explored, but presently, only modified coveralls with attached veils remain the choice of the day. In fact, commercial beekeepers wear these heavy suits as work coveralls to protect from leaking honey, propolis and dirt as much as bee protective suits against stings. As with other beekeeping issues, this heat issue is not insurmountable. You can work beehives in cooler parts of the day or take frequent breaks and drink water.

TIP 47: Rip-stop nylon is a good bee suit material

Rip-stop nylon is a fabric used for parachutes, military clothing and bee protective suits. This style of nylon is very lightweight but dense. Additionally, the bee stinger can span through the nylon but the slick surface seems to make it difficult for bees to grasp the suit material. Rip-stop nylon is available in practically any colour and is lightweight and smooth. It makes for comfortable protection.

TIP 48: White protective clothing isn't an absolute must

It seems that white suits have always been thought to be cooler and that bees were less antagonised by white clothing; therefore, white bee suits continue to be the norm. It may be that beekeepers began using white suits simply because so many other trades wore them and they were simply widely available. However, there are some problems with these. They immediately get dirty and beekeepers really stand out when wearing these full suits. Many years ago, suits made by an English company offered a line of pastel-coloured bee suits. Now, several companies offer pleasantly coloured ones. While the colour of the suit is the beekeeper's choice, bees really do seem antagonised by the colour black. Black gloves, black watch bands, black sweat bands, black eyeglass frames – universally defending bees do not care for the colour. Stick with white or another light colour of your choice.

TIP 49: *For convenience, 'half suits' can be tucked into trousers*

One of my favourite protective suits is a half suit with a veil that zips and is sage-coloured. Though available in white, a beekeeper can choose from several available pastel colours. Most bee work does not require a heavy-duty suit. The half suit can be tucked into trousers so there is no issue with crawling bees bypassing protective gear from the bottom edge of this suit. After work has finished, pull the suit out and unzip it. The suit then has the look of a light jacket, and the attached veil will flip back for taking a breather or getting a drink.

TIP 50: *Expect occasional irritated bees to find small openings*

Bees seemingly have an extraordinary ability to find the smallest opening in any protective gear assembly. No matter that over time, thousands of frustrated bees have been prohibited from getting inside the protective clothing; the few that get through are the ones remembered. Particularly at night when colonies are being moved, the rare bee seems to find the smallest opening. A common place for this small entrance is at the top of the zip fastener. The space initially seems inconsequential, but many beekeepers have had that one bee get inside. A short piece of tape or an attached hook-and-loop strap corrects this small shortage.

TIP 51: *Wear gloves*

🐝 All bee gloves are modified work gloves that have sewn-on gauntlets. More expensive and comfortable bee gloves have a ventilated portion at the wrist to help reduce heat inside the glove. In general, when manipulating hives, bee gloves are necessary evils – clumsy and hot. A common bee glove is the typical reinforced plastic-dipped glove. They are not any better or worse than other types, and sometimes readily accumulate perspiration, causing skin on the hand to shrivel. A beekeeper would rather have these gloves than no gloves at all, but moisture on hands can be distracting and uncomfortable. Several years ago, bee gloves made from supple goat skin were available with gauntlets made from green nylon netting. Compared to other gloves, they were very lightweight. To achieve adequate protection from stinging with these lighter gloves, a long-sleeved shirt had to be worn beneath them. Because they are inexpensive and washable, common canvas gloves are popular. For the money, they are OK, and, if reasonably new, they provide adequate sting protection.

TIP 52: *As your confidence grows, cut back on protective gear*

🐝 As confidence grows and skill improves, it is the rare beekeeper who does not begin to reduce the amount of protective gear that is worn. This is a perfectly normal development sequence. First a lighter suit is substituted for the heavier one, and then, the gloves are worn less and less. After just a couple of years, the beekeeper can quickly get a feel for the mood of a colony on a particular day, allowing the determination of how much protective clothing is required for the hive. But the experienced beekeeper always has the full regalia of protective gear reasonably closely. If something unexpectedly goes wrong, heavier protective clothing is at hand. A veil should always be worn, no matter how quick the inspection is to be.

TIP 53: *Know that veils with pith helmet-styled hats are tricky to balance*

🐝 The new styles of hats that have recently been seen in many apiaries in the UK and Europe are varied. One in particular is the classic styled 'Pith Helmet'. Earlier versions of this had a chin strap, but a helmet without a strap is normal. The problem with this style of veil is that it is very difficult to keep on your head, especially when bending over an open hive or when trying to hold a frame against the sky to have a look inside the cells. It can also be uncomfortable in warm weather.

Mesh design helmets are commercially available and work well until they get crushed. Beekeepers might want to look at other options that do not require a balancing act while working with bees.

TIP 54: *A light fabric veil is great for quick examinations*

🐝 As has been stated elsewhere, a veil should always be worn when a hive is being opened. Fine mesh veils can be wadded and jammed into a pocket and quickly pulled over a hat or cap when needed. However, these inexpensive and convenient veils are not perfect for all situations. Normally, these veils are camouflage, and seeing eggs and larvae through them can be difficult. Since these veils are reasonably priced, they are handy to use as 'back up' veils for beekeepers who do not bring their veil along to an open-hive demonstration. When leading a discussion group around an open hive where some of the other observers are not wearing protective headgear, these lightweight veils can be offered to the individuals for temporary protection. Whether or not they are accepted is up to them. Keeping a mosquito veil in your vehicle will help you to be prepared for an unexpected beekeeping opportunity.

TIP 55: *Keep veil material away from your nose and lips*

When working excited bees, keep the veil material away from your face. The occasional testy bee will sometimes take the brief opportunity to punish you for your intrusion. This is a much more common instance with the light, wispy, fine-mesh veils. However, your eyes will still be protected even if your face is a bit pained. Your nose is a different situation: if it touches the veil, bees can seize the moment. Stings around the eyes, lips and nose are painful. These are the ones that really count. The main purpose of a veil is to protect these sensitive sting spots. You should know that bees specifically target eyes. Exhaled breath directs bees to your nose – a tender spot that often swells when stung.

TIP 56: *A glasses strap is very handy when wearing a veil*

Yes, it's true, a glasses strap is useful, but even so, glasses and veils are an uncomfortable mix – especially bifocal glasses and bee veils. Due to all the extra protective clothing, when the head becomes a bit sweaty, glasses frequently slip and drop into the veil. If wearing gloves, glasses will have to stay off. Additionally, when bending over the hive, perspiration drops will occasionally fall onto glasses lenses, thereby distorting things even more. Wiping them off inside the veil is clearly not an option. Remember that not all days will be hot, sweaty days. Until a grand new advance comes along, the best current recommendation is to wear a veil that allows a light cap or hat underneath – but not the large pith helmet. Use a strap to secure glasses. Be prepared to remove your gloves to reposition glasses to get the best view possible. Most of the time, these precautions are adequate.

THE RIGHT TOOLS

TIP 57: *Buy several hive tools – they are easily misplaced*

Manufactured from spring steel, hive tools are little more than general maintenance tools that have multiple uses around the home. Since bees will seal the hive with propolis, some type of leverage tool is always necessary to open an active beehive. Two common hive tool styles are shown below. Many years ago, some beekeepers must have noted that these tools also had attributes that made them suitable for using in the beehive. It would seem that one tool would be enough to last a beekeeping lifetime. A problem is that they are easily lost in the grass and around the vehicle as hive equipment is moved about. Painting them bright colours helps, but paint will have to be reapplied periodically. As a beekeeper uses this multipurpose tool often, have extra ones ready.

TIP 58: *Explore the different styles of hive tools available*

Years ago, the only real style was the standard hive tool, unlike today where there are many different choices. Some of the newer ones may be helpful for a specific task. The flat, hooked tool, originally made by Maxant Industries in the US, is useful in pulling propolis-coated frames from the box, but it occasionally pulls the top bar off. Try new styles as they develop, and compare them to the traditional tool. There is also a mini hive tool developed for dealing with mating nucleus.

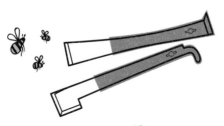

TIP 59: *Be careful of a hive tool in your hip pocket*

Various sheaths have been, and still are, available for stashing the hive tool when not being used. A leatherworker can create a simple carrier that will attach to your belt. If you have one made, leave the bottom unstitched, otherwise propolis and wax will accumulate in the sheath. Most of the time, beekeepers will just drop the tool in a hip pocket. This is a simple and quick method for transporting and the tool is readily available, but you must never forget to remove it before leaving the apiary and heading home.

TIP 60: *Scrape away from yourself*

Nearly every beekeeper has scratched themselves with a hive tool. Though suited for the job, the tool is not perfect. Hive residue can be very tough, and it can be difficult to position the tool to apply the proper shearing force. Invariably, something slips and a cut occurs. No one intentionally injures themselves with a hive tool, but even seasoned beekeepers have experienced this type of accident.

TIP 61: *Tap, rather than push, the tool into position*

When using a hive tool, it's natural to wrap your hand around it – with your thumb on top – to push it into the propolis-sealed crack. Maybe the propolis seal won't break, so you push harder. Suddenly, without warning, everything breaks loose, the tool slips into the opened slot and the first thing that stops this motion is the tip of your thumbnail. It's a very quick procedure, the pain is excruciating and your injured thumbnail takes weeks to grow out. Even though it is tempting to hold the tool in that comfortable fashion, tap the curved end of the tool with the butt of your hand. Lastly, when using the hive tool to scratch out a stinger, be gentle. The heavy sharp tool can cause a greater injury than the stinger ever could.

TIP 62: Have a hammer nearby to make on-the-spot repairs

A common claw hammer should be an integral part of your hive-management tools. If your hives have boxes that are soundly stuck with propolis and wax, use the hammer to tap the hive tool into a sealed joint. If hive nails have begun to back out, nail them back into place. Always have the hammer nearby when dealing with boxes that are encrusted with years of wax and propolis accumulation. Wood frames will occasionally need on-the-spot repair, too. Hive tool use will occasionally pry top bars from end bars. Bottom bars are sometimes stuck down with burr combs and propolis and are pulled off the end bars when pulling stuck frames from propolised colonies.

TIP 63: If frames are soundly stuck, use a hammer here too

The most uncommon use of a hammer as a hive tool is for removing frames that are completely stuck in place from boxes. In many instances, the frame is destroyed before the propolis glue will break. Beekeepers know their colonies and, most of the time, expect this difficulty.

Before the bees get too agitated, try a different tactic. Remove the outer cover and lay the heavy colony – still stuck together with propolis – on its back, and then use the hammer and a standard hive tool to break the seals between boxes. Once these boxes are separated, select a frame near the edge. Gently but firmly tap the bottom bar where it joins with the end bar. It may take a blow or two to loosen the first frame but ultimately you can drive the stuck frame out from the bottom. The frame will need to be driven out with taps on either end. Otherwise, the frame will bind within the hive and become jammed. For more working room, drive out two frames. From that point, the hive box is turned up where the frames can be removed in the traditional way.

TIP 64: *Remove frames gently; otherwise, the top bar will rip off the end bars*

🐝 When significant pressure is applied to the stuck frame with a hive tool, old frames filled with a heavy honeycomb will frequently break or pull apart. Normally, these old frames were only nailed and not glued. Even if they are cross-nailed, the small-diameter 3cm (1¼ in) wire nail will not withstand the pressure. Alternatively, frames that are single-piece plastic have top bars that will flex and allow the hive tool to slip by the frame. The combs in these will be cracked, resulting in leaking honey. Using either the hammer method described above or carefully working the frame with the hive tool, the frame will finally break loose. Yes, this is difficult work and is the reason that the hive should be scraped clean every few years.

TIP 65: *Screwdrivers and paint scrapers make poor substitutes*

🐝 When a hive tool is unavailable, the pliers on a swiss army knife can be misused temporarily to become a hive tool. It's a heavy-duty tool but it is slow to use. Also, it can damage the frames just a bit. Most beekeepers have, on occasion, either forgotten or lost their hive tool. Whatever tool one has in that instance becomes a potential hive tool. Paint scrapers, screwdrivers and lug wrench handles are examples of improvised hive tools. It is as though under the right conditions, 'Any tool can briefly be a hive tool'. Most of these stop-gap tools make miserable hive tools. To avoid using makeshift hive tools, buy more than one and leave the extras beneath the vehicle seat. They are handy tools.

TIP 66: Don't use a hive tool to remove a hive roof

The roof, or outer cover, is a flat cover with edge pieces that extend about 5cm (2in) down on all sides. These side guides allow the beekeeper to quickly replace the cover in the correct position. If this hive roof is used without a crown board, the bees will solidly stick propolis onto the top of the hive. If the beekeeper tries to pry off the outer cover, the main thing that happens is that either the sides or the end rails will break off the outer cover. If you must, give the hive tool a quick, gentle effort. If this effort doesn't work immediately, move on to a hammer. The perfect tool for this job is a short-handled claw hammer. By tapping up on the bottom edge of the hive roof with the hammer, the propolis cracks apart and allows the hive roof to be lifted. When removing or replacing hive roofs, lift and replace using both hands on opposite diagonal corners to give better control.

Do not use heavy pressure on a hive tool to break an outer cover loose from the hive. An inner cover can serve several functions such as an escape board, an isolation device, or a temporary partition within the hive; but its primary function is to protect the hive roof from becoming solidly stuck to the hive. In theory, the crown board can be stuck to the hive roof where the hive tool can easily pierce the sealed cracks and loosen it.

TIP 67: Tools tackle propolis and wax best in cold weather

Propolis and wax shatter much more easily during cold months, resulting in a quicker, easier, cleaner maintenance job. Removing these from the hive in the winter takes less time. At other times of the year, you can use a commercial heat gun to apply high heat to both wax and propolis to soften them enough to clean equipment. It's a slower process but it does work, and the heated propolis and wax will make your shed smell nice.

TIP 68: *Use a second hive tool to clean the first*

It is a rare apiary that has electrical power for operating devices such as a heat gun. Propolis, wax and honey accumulate on the hive tool. Back home, the tool can be properly washed and cleaned with disinfectants, but in the field, disinfecting is more difficult. If in fact more than one tool was purchased, the two tools can be used to scrape each other clean. It's quick and efficient. Bee inspectors must have extremely clean hive tools. Since there is genuine concern about disease spread, tools are scraped often, occasionally with steel wool. A disinfectant may be used or the tool may be briefly heated with alcohol and a flame. Bee inspectors and beekeepers are right to respect disease, and cleaning should frequently be done but it is the rare instance when disease is spread via a contaminated hive tool.

TIP 69: *Store the hive tool together with the smoker*

As equipment is put away, it is not uncommon for some beekeepers to drop the hive tool behind the wire grid that protects the hot barrel of a piece of equipment called a smoker (see tips 80–92) from injuring the beekeeper. In this simple way, the hive tool and the smoker are ready for the next hive inspection. Beekeepers routinely do this. Put the dead smoker, hive tool, hammer, matches, smoker fuel and starter fuel in a heavy, galvanised garbage can with a tight-fitting lid. It's easy to grab and go, and the tight lid helps reduce the smoke odour and prevents a fire from starting.

CONSTRUCTING A HIVE

TIP 70: Try building your own hives (with the correct tools)

Past experience in woodworking skills are a stimulus that nudges many people into becoming beekeepers. It seems that most woodworking beekeepers enjoy building their own equipment for several beekeeping seasons, but increasingly, small-scale hive equipment production can become a chore. As beekeeping abilities grow, often the woodworking beekeeper decides that working with the bees is more rewarding than working in the shop building equipment. But for some beekeepers, the enjoyment lives on. In fact, a few speciality beekeepers develop companies that produce woodenware. Qualified beekeepers should experiment with building some equipment of their own. It is a common aspect of beekeeping that broadens perspective and skill.

TIP 71: For ease, build flat board covers

Building screen board migratory covers is relatively cheap and easy. Simple migratory covers are essentially made of three pieces: a large plywood board at least 1.2cm (½in) thick and two end cleats. The traditional inner cover is completely eliminated when migratory covers are used. These hive tops are named migratory covers because they are actually used in commercial migratory bee operations. They are simple, lightweight covers and since there are no side edges on them, colonies can be packed tightly together and be more stable when loaded onto vehicles. It's not that traditional roofs are impossible to build, but getting a metal sheet, usually either thin tin or aluminium, sizing it, folding the edges, and attaching the shaped metal cover to the frame of the roof is challenging for some.

TIP 72: *Use shanked nails or other such assembly aids*

Equipment made in the home shop can be constructed with care and precision, or equipment can be made as quickly and as inexpensively as possible. Either way, an individual hive can be expected to average about a seven-year life span. Ironically, the bees seem to be just as happy in poorly made equipment as in precision-made equipment. Simple procedures, such as using shanked nails for hive box assembly, can make equipment sturdier. Metal corner braces are also commercially available; these allow quick assembly and provide a strong corner joint. These metal braces can be reused when the time comes to retire the box.

TIP 73: *Try a small block plane for levelling uneven corners*

With either commercially made or homemade equipment, a bit of tweaking is frequently required to bring all edge surfaces in line. A traditional block plane is perfect for this task. The tool has many other woodworking uses, so its purchase can be easily justified. Aligning the surfaces will reduce propolis accumulation and add a neater appearance to the finished equipment – plus it is a quick and easy procedure.

TIP 74: *Assess the pros and cons of buying 'homemade'*

As you would expect, woodworking skills vary from workman to workman. Poorly sized or poorly made equipment is usually not worth the money and effort, but the final decision is an individual one. If you are prepared for the potential inconvenience and the price is low enough, buying rough used equipment could be the right choice. Many experienced beekeepers have a stack of unused or barely usable equipment. Hand-holds are usually the weak feature in such equipment. When making your final decision, consider all aspects of the homemade equipment.

TIP 75: *Use bar clamps to position warped or twisted parts*

Edge surfaces are not the only feature of an unassembled beehive that occasionally needs adjustment. The unassembled components may become warped or twisted after being machined. The defect may be severe enough that, during preliminary assembly, the joints do not align well enough to be nailed in place. A bar clamp or the more common pipe clamp works well for pulling the contrary pieces into alignment long enough for nails to be driven. There are rare pieces that may be resistant enough to pull the joint open a bit even after nailing. Shanked nails or 7.5cm (3in) screws will be better suited for use in this instance. Rub beeswax on screws before driving them into boxes.

TIP 76: *Install metal rabbet liners before assembling boxes*

In an economy line of manufactured hive boxes or boxes made in the home workshop, the wooden lip (the rabbet or rebate joint) on which the top bar ends hang will probably not be covered by a metal strip. However, a better option is the use of a very thin piece of bent tin – approximately 1cm (³⁄₈in) on each side and 37cm (14½in) long – that is tacked into place to protect the wooden lip during future propolis scraping procedures. These simple metal pieces are worth the cost and installation effort. Some beekeepers install them after the box is assembled, but these protective strips are more easily nailed in place on the rabbet joints of the box ends before the box is assembled. The small nails on each far end are particularly difficult to drive once the box is assembled.

TIP 77: Drill pilot holes if wooden components are prone to split

Commonly, soft pine is used for the wide boards used to manufacture beehives. Specifically, Douglas Fir or pitch Pine seem to make a very good beehive. These pines will require that pilot holes be drilled before nailing or screwing together. Particularly, the upper corners of the box ends that are only 1 cm (³⁄₈ in) thick and the side piece into which the nail will be driven will need drilling. When the hive tool is used to break the boxes loose, these are inherently weak parts of the hive that will be exposed to future stress. Nails that are shorter and lighter are included in the kit for this purpose. Wood glue can be used, but in this cross-grain situation it will not have its usual holding power. Red Cedar wood is more expensive but, without doubt, it is used by many beekeepers and is the best you can get.

TIP 78: If possible, assemble equipment on a heavy workbench

A heavy workbench is perfect for hive assembly, but is not an absolute requirement. By design, a good workbench will be of a comfortable height for assembling equipment and the table will be strong enough to absorb the hammer pounding that the hive body will be withstanding. If a stout workbench is not an option, a low comfortable stool on a cement floor will serve just as well. Either way, building a hive is a decent bit of work.

TIP 79: A hand trolley is good for moving bulky loads

A hand trolley or similar cart will be useful in any bee operation. A two-wheeled hand trolley is particularly useful for stacking unassembled hive box parts. After assembly, the cart can be loaded with the assembled boxes to move to storage. In general, nearly any kind of cart or hand truck will occasionally be useful in the bee operation. Hard rubber tyres work better than pneumatic tyres.

WHAT ARE SMOKERS?

TIP 80: *Always fire a smoker when opening a colony*

Cool, white smoke puffed into an open hive disrupts the bees' chemical communication system. This disarming procedure makes the colony look calm when actually the colony is momentarily defenceless. A quick hive inspection without a smoker can be accomplished safely, but there are many variables: the temperament of the bees, the season of the year, even the current weather can all have an effect on the 'quick inspection'. If you are in a rural setting and don't use the smoker, it will take – at high speed – about 5–6 minutes to get a smoker lit if you do end up needing it. So, light the smoker beforehand.

TIP 81: *Thick, white smoke is perfect for subduing feisty bees*

In the beehive, all smoke is not necessarily good smoke. Thick, billowing white smoke is what you want. The thin bluish-white smoke coming from either a hot smoker or an improperly fired smoker will singe bees' wings and cause the bees to become flighty. Plus a smoker producing thin bluish smoke will be very, very hot on all surfaces. Placing a smoker this hot on anything plastic can cause damage or even a fire.

TIP 82: *Be aware, firing a smoker means temporary flames*

All authorities admonish the beekeeper never to allow flames to shoot from the smoker, but during the igniting phase, it is desirable to have copious flames briefly shoot from the open smoker barrel. A coal bed in the bottom of the smoker is what continues to smoulder and produce smoke. If more fuel is added too soon, the smoker may go out just as it is needed.

TIP 83: *Match smoker fuel to the upcoming bee task*

🐝 If you will be working bee colonies for several hours, you will probably want to use slow-burning fuel. Alternatively, a quick inspection can take nothing more than a couple of handfuls of dried grass cuttings or leaves and off you go. There are too many potential fuels to list here. Leaf litter is a quick-burn favourite, while wood shavings and untreated hemp twine are long-burn fuels. Wood shavings need to come from a wood planer rather than a saw. Sawdust can easily smother the developing coal bed. Rotting, wood works great but is not always easy to find. Commercial available twines and paper pulp can be purchased from bee supply companies. Corrugated cardboard can be rolled into tight cylinders and used as a smoke source.

TIP 84: *Avoid fuel that has been chemically treated*

🐝 While smoke can be generated from many fuels, do not try to burn rags that have been chemically treated to be fire-retardant. Rags may look like rags – and hessian sacks are a good option – but those that have been treated to be fire-resistant will show that characteristic long after its usable life is over. Don't burn anything with plastic content as the chemicals released into the atmosphere by burning these are too uncertain.

TIP 85: *Never transport a lighted smoker in an open vehicle*

Smokers can be hard to light and even harder to keep lit, yet if you want them to go out they will seemingly burn for days, particularly in the back of a moving vehicle. There are instances where vehicles and bee equipment have been destroyed due to lit smokers in the back of a moving vehicle producing flames hot enough to ignite nearby fuel. Remember that beeswax is highly flammable and that smoker fuel can be stored somewhere nearby. It's potentially easy to forget the smoker stuck between colonies. A lit smoker producing smoke in such a confined area is enough to leave the van with a smoky smell nearly indefinitely. Use a large bin with a tight-fitting lid. Though it is fireproof, it will readily leak heavy, white smoke into the vehicle.

TIP 86: *Always leave an extinguished smoker outdoors*

There isn't a single good reason to bring a lighted smoker indoors. Even an extinguished smoker will have a strong, lingering odour. Just removing the smoker from the area will eliminate the problem, but the room will still need hours to fully air out. Typically, beekeepers become accustomed to the smoke odour, but to non-beekeepers, the smoke odour is strong. Smokers should be stored outside.

TIP 87: *Improvise a plug for the smoker nozzle*

It is a bit of a bee equipment oddity that no smoker currently being manufactured has a system for closing it off when smoke is no longer needed. Nowhere is there a standard 'plug' for plugging one when smoke is not in immediate need. At a pinch, use a tuft of green grass or, if lucky, you might find a stick of the correct diameter to form a quick, disposable plug. Try making wooden plugs of the appropriate size to plug your smoker nozzle – a wine cork attached to a piece of twine will do. A plug is an extremely temporary object and is easily lost – especially at night when bee colonies are normally moved. The top hole is easily plugged, but the bottom hole is too difficult to access. Generally, it's not worth the effort to plug the bottom tube; therefore, even a plugged smoker can smoulder for hours.

TIP 88: *Extinguish embers before dumping smoker remains*

Clearly, during hot, dry summers a lighted smoker can easily become a fire hazard. This situation is made worse by the fact that often the apiary location does not belong to the beekeeper. A small shovel becomes an important tool – one more piece of equipment that must be carried along – that is required for discarding smoker fuel. It's foolproof to dig a small hole and bury the live embers. Stomping out the embers with our shoes is a common technique, but occasionally after the beekeeper has departed, the smallest ember can recover and become a fire. Your colonies could be destroyed or the land owner's property damaged. Gates left open, damage to driveways and flying bees becoming a pest around equipment seem to concern property owners and other non-beekeepers more than using fire to control bees.

TIP 89: *Keep a smoker handy by holding it between the knees*

This may or may not seem like an issue for some of you, but too much beehive manipulation is 'stoop labour'. Some of you will be wearing heavy protective clothes with heavy gloves and the weather may be warm. Bees are flying all about and you must also manage a hive tool. All the while, the smoker is smouldering. Naturally, setting it on the ground or on a nearby colony is common. However, experienced beekeepers who are anticipating frequent smoking commonly clasp the smoker between their knees. Such a beekeeper does not have to move from the stoop position to retrieve a hot smoker and it is always in easy reach. Obviously, the hot barrel of the smoker cannot be allowed to touch your legs. Additionally, the smoker should have a mature ash bed so it gently smokes while waiting to be stoked for the next use. A smoker just recently lit can emit so much smoke that the beekeeper can't work in the haze.

TIP 90: *When temporarily not in use, lay a smoker on its side*

Smokers are uncanny when it comes to going out. Right when beehive things are firing up and a smoker is needed, it's out. Previously, several suggestions were made about how to prepare a good ash bed and proper fuel, but simply laying the smoker on its side rather than setting it in the normal position may be helpful in squeezing more time out of the fuel load. Heat and smoke rise. Laying the smoker on its side cuts down on that natural draft and reduces the fuel burn rate.

TIP 91: In a pinch, use commercial liquid smoke

Various products that contain the essence of smoke have been marketed for many years. Normally named some variation of 'liquid smoke', these products are the liquefied concentrates of the burning process. The small concentrated sachet is added to 3.8 litres (1 gallon) of water. In some products, spearmint oil is added to make the bees even gentler. Mist the bees, but do not use on uncapped frames. Bees respond reasonably well to this kind of product, but this material is expensive for intensive use. Still, it is not a bad idea to have some around for emergency use. The product seems to last forever.

TIP 92: As far as possible, don't breath the smoke in

The whole notion of smoke use to manage colonies is ingrained in beekeeping history. It is unclear who first had the idea or where the idea was first used. Smoke works well to subdue bees in the hive, but it is not without faults. A perpetual odour lingers after smoking hives, and there are modest concerns about both beekeepers and bees breathing smoke. All types of smoke are not equal. In general, wood smoke is quite similar to cigarette smoke. Individuals with respiratory issues are probably not made better by inhaling bee smoke. There are few bee research studies that have reported the effects of various types of smoke on bees and beekeepers. Indeed, the effects are probably minimal.

For the immediate future, it would seem that the continued use of smoke as a calming agent for bees is assured. There simply is no obvious substitute. Allergies can sometimes flare when burning various kinds of fuels – particularly grass clippings. Sugar syrup sprays are becoming more popular and some beekeepers use these as well, but for the large colony on a hot day with no flow ongoing, nothing can presently beat smoke.

POLLINATION

The partnership between bees and flowers is an old one. As sweet as honey tastes and as much as beekeepers love the responsibility of our charges, at the end .of the day, it is the pollination services that bees provide that is so fundamentally essential to the world in which we live. It's a good idea to learn as much as you can about the importance of pollen, plant life and how to harvest bee nectar.

POLLEN BASICS

TIP 93: *Know the nutritional value of pollen to bees*

All animals, including honeybees, need essential amino acids to produce protein. These building blocks must be obtained externally and cannot be synthesised by bees. Interestingly, honeybees need the same 10 amino acids that humans need for protein production. The only source the bees have for these protein building blocks is pollen. To meet their needs, an average colony, over the entire season, will collect 10–26kg (22–58lb) of pollen. If there is not enough protein, brood rearing slows or even stops.

Adult worker bees will also live shorter lives. Weather conditions such as drought, excessive rain or high temperatures can cause pollen to be in short supply. House bees mix and pack pollen in communal cells that surround the honeybee brood nest. Brightly coloured pollen cells in the brood nest will indicate the blending that bees perform to meet all their nutritional needs. Pollen from different sources has a wide range of colours, from drab to bright primary colours such as red or yellow.

TIP 94: *Appreciate the specialised body hair of forager bees*

Even for humans, collecting and consolidating pollen is a specialised task that requires both labour and equipment. Forager bees are covered in a mat of soft fuzzy hair. Under a microscope, the hair is seen to be branched (plumose), whereas human hair is simply a single shaft. All three of the bee's body segments – head, thorax and abdomen – have abundant hair on them (see diagram, page 87). Even the bees' compound eyes have hair growing from them and the bee's head is covered too. Pollen clings to this hair and is combed off by the grooming bees. These bees mix nectar to make things sticky and then pack the pollen into specialised leg pouches.

TIP 95: *Electrostatic charges attract pollen to forager bees*

As a bee flies, to a greater or lesser extent, she acquires a positive electrical charge. When she approaches a blossom that is attached to a twig, which is subsequently electrically grounded through the tree, a static electrical discharge causes some species of pollen to literally fly up to meet the approaching bee. This action improves pollination efficiency and allows the bee to work more quickly.

TIP 96: *Pollen from wind-pollinated plants is important*

Plants have evolved several different pollen distribution systems. Two common ones that are involved in beekeeping are wind-pollinated plants and insect-pollinated plants.

Frequently, tree pollen produces wind-borne pollen. Additionally, corn is a common wind-pollinated plant from which bees will collect pollen. Apparently, this wind-distributed pollen does not meet all the amino acid needs of the colony, but, if mixed with other more bee-friendly pollens, these sources do seem beneficial to bees. In fact, maple tree pollen is a significant early pollen source for a colony coming out of winter.

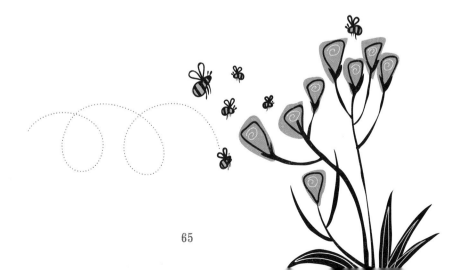

TIP 97: *Expect reports of your bees visiting other feed sites*

🐝 In early spring, bees are frantic for the first sources of pollen. Unlike honey, pollen is not stored for long periods and reserves are quickly used to nurture developing brood in the early spring. Early-season foragers are common visitors at feed bunkers, where they will rummage through animal feed. Possibly, some added syrup will be attractive to them. At bird feeders, forager bees will scrabble through the bird seed dust. Oddly, desperate early-season honeybees have been seen trying to remove paint from farm machinery. In such cases, the foragers have little success in gathering, packing and transferring the material back to the hive. As soon as any natural source becomes available, bees will stop this aberrant foraging and begin to work normally.

TIP 98: *Use pollen traps to harvest pollen when scarce*

🐝 As a bee enters the hive, pollen traps strip pollen pellets from her pollen basket. (See tip 101 for pollen trap uses.) Commonly, bees are required to pass through galvanised solid grids having numerous holes measuring 5mm ($^3/_{16}$–$^7/_{32}$ in) in diameter. Such a grid just barely allows a bee to squeeze through the opening. This narrow entrance dislodges the pollen pellets on the forager's rear legs.

When traps are in place, most pollen pellets do not get to the colony's brood nest, causing the colony to begin experiencing a pollen shortage and the worker population to decline. Some models of these devices can be made in a home workshop, while other, more complex, models are available commercially. Neither one – home nor commercial – is liked by foraging bees. All traps must make allowances for drones, which are entirely too large to squeeze through the small entrances. One-way entrances and otherwise small entrances generally allow the drones to leave the colony.

TIP 99: ...But know that traps will stress the colony

🐝 Pollen traps are essentially entrance-obstructing devices that all bees seem to dislike. The brood nest bees will become aware of the developing protein shortage. Entrance congestion will result from workers and drones trying to return. Over time, there will be a movement towards gathering pollen from plants that result in smaller pollen loads that get through the grids more often. Recognising the disruption caused to the colony, many trap designs have a system that allows the beekeeper to manually bypass the collection grids. Commonly, about every third day, the collection system is bypassed to allow the colony a few days of unabated pollen collection.

TIP 100: Stored pollen isn't great for pollination procedures

🐝 In some instances, pollen is collected to be used in either hand- or mechanised pollination procedures. Manual pollination is accomplished with a camel-hair brush or with a mechanical blower gun. Using a small, dampened artist brush, pollen is moved from one flower to another. The process is tedious. Blower guns use harvested pollen that a person can aim and put a puff of pollen on a blossom. The blower gun is much faster, but requires harvested pollen.

Collecting the pollen is labour-intensive and only used in high-value situations. Honeybee-collected pollen would seem to be the answer to this challenge, but in fact it isn't. House bees mix pollen with glandular secretions to produce 'bee bread' to serve as a protein food for young bees. These salivary enzymes begin the process of digesting the pollen grains' outer coat, thereby rendering them useless for germination.

TIP 101: *Mix bee-collected pollen with protein supplements*

Trapped pollen can be used either as human food or mixed with protein supplements to form a 'pollen extender'. This artificial protein is then fed back to the bees as they need it. However, all pollen must be from disease-free colonies or sterilised in some way. In some areas, due to a fear of disease spread, this re-feeding procedure may not be allowed. However, in most instances, a beekeeper can collect pollen from personal bee colonies and then feed natural pollen back to personal bee colonies mixed with powdered protein supplements. Even a small amount of pollen in a protein supplement makes it more attractive to the needy colony.

TIP 102: *Cold-store pollen in small batches*

Most traps will inadvertently mix bee parts and other trash with the collected pollen. Pollen must be manually cleaned and dried under low heat from a source such as an incandescent bulb. Freeze it for future use. Due to bacterial activity that occurs in the core of the stored pollen, freeze pollen in small batches of 1.8–2.2kg (4–5lb). Do not store in large containers.

WHAT IS NECTAR FLOW?

TIP 103: *A beekeeper can't provide a true nectar flow*

⬢ A nectar flow is the time of the year when bees make a honey crop. It occurs during the warm months when flowering plants are producing nectar for pollinators. Most major nectar flows are in mid-spring. People who don't understand the biology of honeybees and honey production often ask if a beekeeper can simply feed some table sugar to bees to make a honey crop. The answer is 'technically, yes' but 'realistically, no'. Sucrose is a common sugar that bees enzymatically reduce to the simpler sugars glucose (dextrose) and fructose (levulose).

If a colony is near starvation or if the beekeeper feels the colony needs stimulating, sugar is fed to the bees and they convert it into energy and young bees. Feeding enough sucrose (table sugar) to produce a meaningful nectar flow would be a major undertaking. Possibly, commercially purchased bee feeding syrup would be easier to feed in quantity, but either feed will result in a bland, near-colourless product. Providing a meaningful nectar flow is not realistic.

TIP 104: *Grasp that nectar is only modified plant sap*

⬢ Without nectar, there would be no honey. Nectar is a thin, watery, faintly sweet liquid. Bees simplify the sugars and remove most of the water to make it into honey. The nectar from which honey is made is basically plant sap that has undergone some chemical changes during the secretory process.

TIP 105: *Nectar does not always come from blossoms*

✿ Extrafloral nectaries differ from floral nectaries in that no obvious flower structure is apparent. Many plants have these cellular structures on different parts of the plant. Salix, beans, prunus and lime trees are examples of plants having extrafloral nectaries that have been studied. Honeybees will readily visit these non-pollination plant sites. Apparently, the primary purpose of such plant structures is to attract ants that will help defend the plant from other, more damaging, pests. On occasion, a beekeeper may observe a mysterious nectar flow that is being provided by these hidden nectaries.

TIP 106: *Worry not about how to tell if the flow is ongoing*

✿ Novice beekeepers often ask how to tell when a flow is on and in what ways preparation should be made for it. Historically, local beekeepers will know the average dates that primary flows begin and end. Watch the progression of the primary crop blossom for a clue. But when the nectar flow begins in earnest, it will be obvious. New honeycombs will appear nearly overnight. Thin, watery nectar will be in any available comb. As the combs are examined, thin nectar will dribble on bee suits and shoes.

The bees will also be noticeably manageable during nectar flow. The reasons for the bees being agreeable are not clear, but possibly simple statistics may play a role. So many bees are out foraging that there are fewer bees to bother the beekeeper. Additionally, with so much food available, robber bees will not be an issue. Bees robbing each other is a significant cause for excessive colony defense.

TIP 107: *Nectar production is needed by bees and plants*

🐝 Nectar is a type of bribe that plants offer to foraging bees to achieve pollination services but, alternatively, it is more of a mutual product needed by both bees and plants. Bees need it as a carbohydrate food source, while plants need it to attract bees for pollination. The plant needs to supply just enough to keep the forager interested in the reward – but not too much. It's a complex relationship.

TIP 108: *Pollinated blossoms quickly stop producing nectar*

🐝 Once a blossom is pollinated, the pollination game for that flower is over. Quickly, the plant folds up shop and redirects its energy to other still-developing blossoms. This blossom count decline is a useful observation for beekeepers. As more and more blossoms undergo this procedure, the nectar source peaks for the season and begins to decline. Other sources must be sought immediately.

Other than age and successful pollination, other things that stop blossoms from producing are: late season frosts, violent weather such as summer thunderstorms, mowing or harvesting or drought. Canola (oil seed rape) are normally harvested, and apple trees are normally sprayed, shortly after they flower. This sudden loss of nectar source can leave a colony with an uncapped honey crop and no incoming nectar from which to produce wax.
Ideally, another crop will be found.

TIP 109: *Nectar-collecting eagerness varies among colonies*

There is still much that is unknown about honeybee foraging behaviour. The bees' ability to find and exploit nectar sources is uncanny. Many variables must be considered, but the main ones are: genetics, weather, availability of floral sources and storage space.

Since the earliest days of apiculture, a major goal has been to select for strains of honeybees that are high honey hoarders. Today's bees will put up far more honey stores than they will ever need during winter seasons. But this savvy foraging behaviour is not universal in every apiary. Some colonies are better foragers that others – and not just for nectar. Foraging for pollen is also variable and, obviously, defensive behaviour varies. Some colonies are just hotter than others. But in general, expect a range of surplus honey crops in the apiary.

TIP 110: *Expect bees to forage far from their home apiary*

A very common question from both beekeepers and property owners is, 'What crops should I plant for bees?' In most cases, the question has a simple answer – 'Grow plants that you like and that please you'. Bees from a single colony will on average, forage about 3 miles (5km) in any direction – that radius equals just about 18,000 acres (see tip 159). It logically follows that the closer the food source, the more efficient the foragers will be at harvesting it. If they have to fly miles away, the food reward will become increasingly small.

TIP 111: *Big honey crops require big acreages*

🐝 The typical crops that yield the large honey crops are well known. Some common examples are: clover, canola, heather and apples. If the area is residential and small gardens are maintained in most house lots, then the small planting contributed by each homeowner becomes substantial to the beehive – especially since garden crops are watered and fertilised.

TIP 112: *Provide abundant hive space during the flow*

🐝 Empty supers are the storage space that timely beekeepers provide for a colony during a nectar flow. Ideally, more space is provided than is needed. A good rule is for the top super to always be nearly empty. In that way, no part of the crop is ever lost by the bees and the beekeeper (see beehive diagram, page 33). However, if the inner cover is solidly stuck to the top super and burr combs abound, some part of the crop is lost. A little-known reason for extra comb space is that bees need this to temporarily spread thin nectar for evaporation purposes. As the nectar season for a particular crop peaks, the beekeeper should slightly under-super to encourage bees to finish capping the crop they already have.

POLLINATORS IN ACTION

TIP 113: *Insect pollination shouldn't be taken for granted*

The common statement: 'Our food supply could not survive without insect pollination' has been made time and again, and it is very true. But insect pollinators have never let us down – not once. As food consumers, we just always expect them and the food supply to be available. Even though pollinator populations are declining, there has never been a serious shortage of pollinated foodstuffs due to failed pollination. Even though bees have always provided pollination services, and even though the importance of these services is well known, the bees' hard work is often taken for granted. Insect pollination is truly important to our way of life.

TIP 114: *Nectar and pollen reward per blossom is very small*

The plant species that requires insect pollination is faced with a basic challenge. It must supply nectar and pollen – or pollen alone – in just the right small amount – no more, no less. If the blossoms of a particular plant supply copious amounts of nectar e.g. *Brassica napus* (oil seed rape) or heather, foragers can get all they need from one blossom and will not need to move from blossom to blossom to cross-pollinate. However, if the plant does not provide enough nectar to keep the forager's interest, the bee may switch to a species with a more liberal reward. Additionally, the plant energy used to secrete nectar cannot be used for plant growth or seed production. Indeed, in some plants, nectar is not used as a reward. Pollen production without nectar production is even more complicated. Nectar and pollen production rates are an interestingly complicated aspect of pollination.

TIP 115: Only expect a few bees on small plantings

It is logical that only a few bees will be attracted to a small planting. There is simply not enough reward for large numbers of bees to compete for small returns. Additionally, other pollinators may also be vying for those same meager rewards.

TIP 116: Know that insects are excellent pollinators

There are about 270 species of bee in the UK compared with 3000 species of bees in the US. Many settlers from Europe and the UK tried in vain to take honeybees on the long journey to the 'New World' (America) but with disastrous consequences. In and around 1622, early pilgrim settlers did manage to introduce honeybees (*Apis mellifera*). In addition to honey and wax, bees also provided pollination services for crops. It was believed that honeybees were not native to America, however recently, fossilized evidence of the honeybees *Apis Caspacapis nearctica* from the Miocene period of Nevada have been found. This was therefore proof that there were honeybees in the US but they had become extinct.

Native pollinators will frequently work nearer to the home nest and will forage on smaller blossom populations. In some cases, plants are better suited for a specific pollinator. Honeybees are good generalists and exist in very large numbers and generally get the job done on any plant. Many native bees are specialists, and on selected plants may be considered to be excellent pollinators. None of these native bees produce surplus honey crops. Though they may not have the same reputation as honeybees, they are valued pollinators that handle numerous pollination tasks.

TIP 117: *Large acreages of insect-pollinated crops will require extra bees*

Commercial farming is commonly achieved by cultivating large acreages of crops. Some fields are hundreds of acres. Some of those crops, such as beans and sunflower, require insect pollination to produce commercial seed sets. Too often, foragers can get all the necessary food supplies that they need near the edges of the planting, so pollination activities go lacking near the centre of such large fields. Additionally, it cannot be assumed that all foragers will stay on the targeted crop. In general, large plantings require large foraging populations placed throughout.

TIP 118: *Bees need more than a large monoculture*

Large monoculture plantings can stress commercial honeybee colonies by not providing a balanced food supply. For example, this nutritional shortage sometimes occurs in colonies used in large-scale almond pollination where almond pollen is the primary protein source. Other large-scale plantings, like cucumbers or watermelons, have low flower densities per acre. Pollination colonies on those plants may decline because of food supplies.

76

TIP 119: *Fields or orchards are great places for bees to roam*

🌳 Modern fields and orchards are not natural settings for bees. In many cases, such crops cover many acres and will contain only one type of flowering plant. Honeybee colonies are commonly brought in to these plantings to supplement the natural pollinator activity. Foraging bees will partition the surrounding resources and forage as efficiently as possible. There is no practical reason for a forager to fly a great distance from the colony if their demand can be fulfilled much closer. In general, in commercial pollination settings (see tip 24), the foragers will stay within a few hundred yards of the hive. Though much easier for the beekeeper, setting off large numbers of hives in a block will not result in equal pollination activity; therefore, beekeepers will set clumps of hives equidistant around and within the planting to be sure that all parts of the crop are exposed to bee forager pollination activities.

TIP 120: *Only use strong populous colonies for pollination*

🐝 A primary stimulus necessary for a colony to send foragers to gather pollen is open brood. Colonies do routinely store some pollen for future use, but when populations are expanding rapidly, pollen reserves can be quickly depleted. Ideally, colonies to be used in commercial pollination activities should be healthy, populous and have a high population of open brood. Regardless of open brood percentages, all healthy colonies will forage for pollen, but colonies with a high percentage of open brood will be better.

TIP 121: Don't fret about pollen and nectar coming from competing crops

Within fruit orchards, growers often spray on various herbicides to kill flowering plants like dandelion and clover. The grower's intent is to remove competing nectar and pollen crops so that more pollination attention will be given to the target crop. While no great harm is done in this procedure, there is probably little grower reward in the effort. Unless a vast effort is undertaken on all surrounding lands, wayward forager bees will just travel a bit further to find another attractive source. Bees seem to instinctively know that any nectar and pollen source is temporary and will soon begin to wane. In order to stay a step ahead, some bee foragers are always searching for alternative food sources in order to quickly replace the current one. Don't worry too much about competing crops within the orchard. The efforts to remove them are not worth the reward.

TIP 122: Realise that pollination services have to be precise

In the UK commercial pollination – even on a small scale – is a time-sensitive endeavour. Regardless of rain or temperatures, colonies must be moved in on the average date of bloom. A few days later, those same colonies must be moved out on a tight schedule in order for crop production programmes to continue on schedule. Commercial pollinator equipment must be sound and dependable. Alternatively, large pollination operations may have specialised hive loaders. Occasionally, for smaller commercial pollination operations, trailers are used to quickly drop off colonies and then quickly retrieve them when the call comes. In special instances, the grower will provide trailers for the beekeeper to use temporarily. But much of the time, smaller numbers of colonies are simply hauled in a large van and manually unloaded.

POLLINATION AND YOUR GARDEN

TIP 123: *Foraging bees on blossoms are not prone to sting*

While foraging bees can surely sting, they rarely will unless handled or crushed. Broadly, honeybees are inclined to sting for two reasons: ❶ to protect the nest and hive stores and ❷ to defend the species (see tips 162–166). At any given time, probably 1–2% (a few hundred bees) of the colony bees are guard bees. If the nest is threatened, they will begin stinging to protect the colony. On flowers, foragers are more apt to fly away than attempt to sting.

TIP 124: *Beehives should be near, not in, the garden*

There may be aesthetic reasons for putting a colony specifically in the garden, but from a forager's perspective, roaming bees will find flowering garden plants even if they are separated by hundreds of feet. (See tips 188–200 on hive placement.) For gardeners working near the hive, there is a risk of an exiting or returning bee accidentally dropping down a shirt collar or becoming entangled in hair. A sting will probably result. If possible, stay out of the bees' flight path.

TIP 125: Small gardens provide small but good rewards

Even if a beehive is positioned nearby, a small garden will attract only small numbers of pollinators. More frequently this is an issue in small pollination plots where a grower pays for pollination services but it appears that most bees are flying far away. In fact, they probably are. Rewards from single blossoms are limited and stop when pollination occurs. Don't expect a lot of bee activity in a small garden or a significant honey crop from one. Small gardens provide small rewards for bees, but pollinated gardens provide wonderful produce for the gardener.

TIP 126: Weeds can be an excellent food resource

Weeds are common nuisances for gardens, lawns and crops. Incredible amounts of money are spent to suppress weed growth. Yet to foraging bees, weeds are perfectly acceptable as food sources. All of the interaction between blossoms and bees is in play even if the blossoms are on 'weed' plants. Asking growers and homeowners not to suppress weed growth is impractical and unrealistic. Maybe a more rational request would be to ask that not all weeds be killed but only the ones that are truly disruptive. Neighbouring homeowners who are not beekeepers frequently have different opinions about allowing clover growth in lawns or weed growth areas on the back of the property. Know that weeds can be excellent food sources for bees, but also that they are not welcomed in most crop, garden and lawn environments. Also remember that unrestrained weed growth generally frustrates neighbours.

TIP 127: *Remember: two hives are better than one*

🐝 The small garden pollinator should ideally keep two colonies rather than one – especially if the gardener is new to beekeeping. If something goes wrong, the gardening beekeeper with a single colony has less recourse to a solution. If two colonies are maintained, resources from one could be used to help the other. This tip is not necessary if the gardening beekeeper has other beekeeping friends who can offer assistance in times of colony need. If it is practical, keep two colonies.

TIP 128: *Realistically plan your garden preparation time*

🐝 Assembling and painting the hive in your garden, and then installing bees will take something like 10–20 hours. Beyond that, for one or two hives, an hour or two per week would be an average time required to manage the colonies (see chapter 5).

TIP 129: *Use pesticides, but with caution*

🐝 Depending on the gardener's viewpoint, pesticides may or may not be used. It is an individual decision (see chapter 7). When applying dusts or spraying liquid insecticides, watch for pesticides drifting towards the colonies. Spraying at dawn or dusk is helpful. Winds are usually quietest then. Be aware that water sources may become contaminated with pesticides. At times, insecticides are helpful, but be aware that unintended consequences can result from applying these compounds when bees are near.

TIP 130: *Water sources must be kept clean and fresh*

🐝 As the hive interior heats, young bees in the brood nest will increasingly require water to cool the brood nest. During summer months, water-foraging bees will visit bird baths, dripping water faucets, nearby swimming pools, ponds – anywhere there is dependable water. If rain has been abundant, water foragers will gather water from standing pools. Water demands are most obvious during the hot months. Be certain that a water source is somewhere near the garden and that it is kept fresh and clean.

TIP 131: Honeybees are not great pollinators for greenhouse crops

Greenhouse plants that require pollination, such as greenhouse tomatoes, must be pollinated. Tomatoes can be hand-pollinated, but that task is laborious. Honeybees have been tried and do accomplish the task, but human workers are commonly uncomfortable with so many bees buzzing around the greenhouse. Most of these bees become lost and die on the greenhouse floor. Bumblebees have been found to do an admirable pollination job in confined areas such as greenhouses. The bumblebee nests are smaller with a smaller forager population. Additionally, they seem to be more at ease with the confined environment. If bumbles are not available, honeybees will work, but not as well.

TIP 132: You can't prevent bees from foraging on invasive plants

To many beekeepers, it is concerning that their bees happily forage on plants that are thought to be noxious and invasive. Indeed through the years, unknowing beekeepers have been responsible for distributing seeds or plants before it was understood that these plants would become invasive weeds. Two such plants that are good nectar and pollen plants but otherwise invasive plants are Japanese Knotweed (*Fallopia Japonica*) or Common or Great Rag Weed (*Ambrosia artemisiifolia* and *Ambrosia trifida*). Initially, both of these plants seem to have merit. Each is colourful and hardy and bees love them. Even though bees thrive on them, these plants and other invasive plants like them should not be encouraged. Review the noxious weed list for your area and know what plants to avoid before beginning to plant. With a little research, a gardener can avoid propagating uncontainable plantings and provide plants that have value for bees.

HONEYBEE BIOLOGY
AND BEHAVIOUR

A fundamental understanding of bee biology and behaviour is a basic requirement for becoming an accomplished beekeeper. Gaining bee knowledge has no obvious endpoint. New information about bees and their ways is constantly evolving, requiring the beekeeper to constantly stay informed and updated. This ever-expanding knowledge pool will help attentive beekeepers become better at caring for their bees.

UNDERSTANDING BEES

TIP 133: *Know the biology of a bee and colony roles*

🐝 The three types of bees within a colony – workers, queen and drones – have major body parts in common, but smaller or specialised body part functions may vary. For instance, a drone does not have a sting structure or wax glands, but does have male reproductive parts. A queen has atrophied wax glands, but has a fully functional reproductive system. Workers only rarely produce a few infertile eggs. All three honeybee caste members are noticeably similar, but each is different in important aspects.

WORKERS: Worker bees are by far the most commonly seen honeybee caste member. As their name indicates, worker bees; which are reproductively sterile, perform most of the hive tasks, such as: building and maintaining combs, feeding brood, incubating brood, foraging for colony resources and defending their colony when necessary. During warm months, workers number in the tens of thousands within the colony.

QUEEN: The lone queen is responsible for producing thousands of eggs that develop into replacement bees. Additionally, her glandular system produces pheromones that inform and stabilise the colony's various behaviours. The queen lives for a year or two before being replaced. There is rarely more than one mated queen within the colony.

DRONES: Drones are specialised bees that are known as haploid as they have 50% of the genetic complement of queens and worker bees. Their function is to breed with an unmated queen high in the air; drones are rarely seen at ground level. Due to the rarity of an unmated queen, most drones do not copulate successfully. Drones are only present during warm seasons and number about 300–600 per healthy colony.

The body of a worker bee

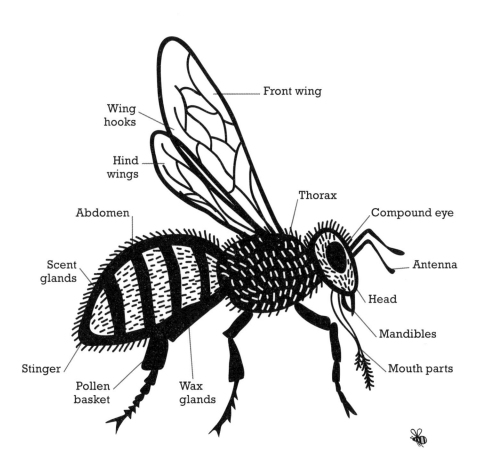

Front wing

Wing hooks

Hind wings

Abdomen

Scent glands

Thorax

Compound eye

Antenna

Head

Mandibles

Stinger

Pollen basket

Wax glands

Mouth parts

BEE COLONIES AND THEIR NESTS

TIP 134: *A bee colony has a clear seasonal cycle*

It is a common misconception that a bee colony grows to a mature size and stays at that point. That is not the case and for good reason. Bees must make preparation for long seasons without any external food resources coming into the nest (for more about the beekeeping year, see chapter 8). When times are good they must store for lean periods. Bees build high populations to collect food during these bountiful times, but when times are lean and only stored food is available, the adult population is cut dramatically. This seasonal cycle of population ups and downs is a logical procedure for the 'feast-to-famine' lifestyle of the honeybee.

TIP 135: *Different colonies will have different personalities*

A good, experienced beekeeper will know the personality of his/her hives. In general, some colonies are enjoyable to open, while others are simply miserable to manipulate. Even more confusing is that the colony's personality is highly variable. In the spring season, colonies are much easier to handle than in the hot summer when most nectar and pollen sources are over for the season. Essentially, bees are in good moods when resources are plentiful – but behavioural genetics are important in dictating how the bees will respond to an intrusion into their colony. Even in times of resource abundance, some colonies are much more defensive than other similar colonies at the same time and same place.

TIP 136: Humans will never totally understand bees, but we must try

Throughout humans' experience with honeybees, we have had to struggle not to impose our mannerisms and needs on the honeybee. Indeed, we have named the primary egg layer the 'queen' and all the smaller bees we labelled 'workers'. Beekeepers paint the bees' hives nice colours and provide a front porch (a landing board). We feel sorry for the drones during the autumn drone massacre, and we fret when the queen must be killed before introducing a new queen. Maurice Maeterlink, Nobel Prize winner for literature, sums this all up in his book entitled *The Life of the Bee*, 1901.

TIP 137: Remember, bees live in total darkness

Nearly all photos and video productions of honeybees are produced in clear bright light. Yet, inside the hive, visible light is essentially nonexistent. The inside of a healthy, populous beehive in warm weather is pitch dark, hot and very crowded. Honeybees live their lives in two worlds – one of total darkness and the other outside in full sunlight. Apparently bees need two different navigation systems. Inside the hive they probably rely on cues like odours, gravity, electrical fields and touch. Outside the hive the bees have a different navigational system that uses the visible light spectrum (except for red) and allows them to see into the ultraviolet spectrum. Beekeepers should remember that the way they see the inside of a hive is not how it looks to bees once the cover is back on it.

TIP 138: *Know that bees won't always select great nest sites*

Bees do often make mistakes – even fatal ones. While the artificial nest that beekeepers have developed is not perfect for the bees, often proper nest cavities are simply not available. Scout bees are forced to select an improbable place such as beneath a barbecue grill or underneath the lid of a partially opened dustbin. While these nest cavities are far short of ideal, they are out of the weather. An exposed nest is the biggest error. Sometimes, the swarm leaves and pitches in a heavily shaded location. Upon finding no other nest cavity, the bees begin to construct combs in what is, at the time, a cavity-like structure. However, autumn approaches and the leaves drop and the nest will become exposed. This nest will die even in a mild winter. The beekeeper should expect bees to select poor nest locations.

TIP 139: *...And all-natural nests are hardly ever textbook perfect, either*

Inside the natural nest, depending on the nest cavity, combs will weave and twist on each other. If the nest area is small, the colony will outgrow it and swarm. Even when bees find a nest cavity, it is not always one that will allow the colony to thrive. The success and productivity of natural nests will vary. Living in the wild is not always great for honeybees.

TIP 140: *Understand the concept of 'bee space'*

🐝 Though at times difficult to see in use, bee space is used throughout the natural nest. Bee space is essentially the space required for two bees to squeeze their thoraces by each other – back-to-back. That space averages 6mm–1cm (¼in–³⁄₈in). That is the space needed for two bees to pass each other within the colony. Spaces larger than 1cm (³⁄₈in) wide can be used for storage space. The average nest cavity is about one cubic foot. A productive colony can quickly fill that area, so in good years no space is wasted. Spaces less than 6mm (¼in) are inaccessible to the bees but are readily accessible to early-stage wax moth larvae and all stages of the small hive beetle, so these spaces are sealed up with the hive caulking compound – propolis. While bee space dimensions may appear fixed and rigid, some colonies will vary a bit depending on hive cavity layout and bee population. Bee space is always used in the nest, but the actual space is a bit variable.

TIP 141: *Bees might not build a nectar flow comb*

🐝 Wax is an expensive product for members of the bee nest to produce. Huge numbers are required to estimate and explain the beeswax secretion process. Bees that are 12–18 days old are physiologically responsible for producing beeswax scales (each scale about the size of a pinhead) from four pairs of glands on the bottom side of the bee's abdomen. It takes about 800,000 scales to produce about 0.45kg (1lb) of beeswax. Bees must visit about 17 million flowers to gather the 3.8kg (8½lb) of honey needed to secrete 0.45kg (1lb) of wax, which requires flying about 753,170km (468,000 miles) (approximately to the moon and back). Wax secretion for comb construction only begins when all space in the hive is filled and the bees' internal storage spaces are filled. At that point, select house bees' wax glands begin to produce wax scales which are moulded into combs. Without a nectar flow or additional space, bees will not build new combs.

TIP 142: *Construction bees will make their own foundations*

🐝 On occasion, beekeepers do not supply comb foundation (sometimes called foundation sheets) for a colony. If no frames are supplied, the colony will build natural, freestyle combs. But sometimes frames are in place without foundation. This absence is commonly due to the beekeeper's oversight or error. But even if foundationless frames are present, bees will not readily acknowledge their orientation and will build combs to their natural liking – which is unlikely to be to the average beekeeper's liking. Don't forget foundation or, if you do, expect a mess!

TIP 143: *Brace comb construction will happen post-move*

🐝 Little is known about how the bee nest in a tree cavity deals with the wind-flexing of the nest combs. But based on observations made with managed colonies, it would appear that bees do not like combs swaying and bumping each other. Frequently, after a colony has been moved from one location to another, it appears that additional comb is constructed in unwanted areas for the beekeeper. 'Brace' and 'burr' combs are the popular names, while others include 'bridging' and 'ladder' combs. Brace combs attach combs to the cavity walls and to each other, while ladder combs are exactly that, ladders.

Ladder combs may be seen on the bottom boards, where it is assumed bees use them to get to the frame above. Comb 'globs' may also be seen on frame top bars to provide access to the combs above. The bees seem to be fine without these specialised combs, and some colonies seem to build them more than others. Maybe these uses are not the intent, but they do seem to be built more in colonies that have been recently moved and in colonies not routinely managed.

TIP 144: *New, white comb is easily damaged*

🐝 The beeswax scale that is the building block of new beeswax combs is secreted as tasteless, odourless and opaque white. New beeswax combs are delicate and pliable. New combs are easily damaged, but seemingly easily repaired by house bees. Bees initially reinforce the fragile combs with propolis rims. Over time, they add thin propolis films to the cell walls. The combs become much more rigid and acquire the typical yellow colour of new wax. Apparently the colour comes from pollen and propolis, while the characteristic odour of wax is thought to have come primarily from propolis. Fragile, 'virgin' beeswax generally only exists for a brief time during the strong spring nectar flow.

In some areas, there may be a meaningful autumn flow that could result in comb construction, but generally, the autumn flow is less intense and new combs are not eagerly constructed. Be careful when handling frames of new, white combs, they can easily tear and break.

THE ROLE OF QUEEN BEES

TIP 145: *Egg production can tell you a lot*

Seeing eggs in your colony's brood nest area is almost as good as actually seeing the queen. Unless there is a severe physical problem, you can't tell much about the queen just by looking at her. Size alone does not indicate a great queen, and colour is strictly a matter of choice. In most cases, keeping the colony opened for longer and removing more frames just to see the queen is needlessly disruptive. In instances where a new queen was just introduced and does not yet have a large brood nest established, just seeing eggs is enough to know that she was there at least three days ago. That is normally enough to suspect that the queen introduction procedure has worked acceptably.

TIP 146: *Evaluating the brood nest is a helpful guide*

How many eggs are present in relation to the time of the year is important. During your first inspection, if there are eggs throughout the brood box, that would be the sign of an on-the-job queen. If egg numbers are scant or spotty, or even nonexistent, that indicates a problem for you and for that hive. A good, productive queen, heading a legion of eager, instinctively alert brood bees, should have a concisely organised brood nest that is developing in a series of concentric circles much like a bull's-eye target. Brood nests that are disorganised and scattered – especially if the queen has already put brood in drone cells – indicates a colony that is in need of a new queen or may be subject to disease.

TIP 147: 'Good' queens can perform poorly

🐝 Large, nicely shaped queens can still be poor producers of eggs. This tip can be broadened to read, 'Watch for any queen performing poorly'. But the point is that even young, healthy-looking queens are not guaranteed to produce vibrant, productive workers. Alternatively, a smallish, misshapen queen may produce a great worker population. No queens come with a guarantee, so the beekeeper should always be observant and constantly be making decisions on the suitability of the present queen.

TIP 148: A colony's queen should be replaced

🐝 It is not uncommon for beekeepers to keep a queen longer than they should. This reticence to replace her might have some anthropomorphic connection with humans growing old in a particular job and being summarily replaced. If the beekeeper does not replace the ailing queen, the bees will do it, and, they are frequently much less tender in their methods. Too often a compassionate beekeeper will put a failing queen in a cage, with the best of intentions to use her in an observation hive or some such use, but many times, the queen is forgotten and dies a slower death than necessary.

Queens today seem to live productively for only about a year. Even though distasteful, when the time comes to replace a queen, the most efficient procedure is to quickly kill her. Beekeepers tend to refer to this method as 'snipping the queen' or 'pinching between the ears'.

TIP 149: *Note that nurse bees are the colony nursery keepers*

The queen is not the colony leader but is more like a sophisticated piece of specialised equipment. She is the genetic reservoir for the colony and she is an egg-laying device. A select group of seemingly unremarkable young brood nest bees actually run the show. These nursery keepers prepare cells – drone, worker and occasionally queen cells – they feed the queen the appropriate rations for the season, and they instruct foragers to bring water, nectar and/or pollen. If the queen makes a mistake and puts a drone egg in a worker cell, this cadre of young workers eats the egg and prepares that cell once again for the queen. This transient group of workers keeps the brood nest orderly and on schedule.

The queen is maintained and supported, but she is not a decision-maker for the colony. In just a few days, these nurse bees will grow into other stages of their lives and become supporters of the brood nest workers. The queen always gets the credit, but she does not do the job.

TIP 150: Some drone production can be healthy

As recently as 1975, it was thought that high drone populations were detrimental to the honey production of the colony and the general recommendation was to destroy drones and drone brood whenever possible. That recommendation was totally incorrect. The healthy bee nest will have about 13–17% drone comb and, throughout a typical season, that colony will produce 5,000–20,000 drones. Only about 50–200 of these drones will be successful in mating with a queen, but the genetic reward is so great that the expenditure of resources is worth it to the colony. The colony members are aware of the presence of drones via chemical messengers. When beekeepers destroyed drone and drone combs, colonies would frantically rebuild to the desired level of drone population. It was a waste of time for the beekeeper and a waste of energy and resources for the colony. Beekeepers should accept a population of 400–600 drones as a sign of a healthy colony.

TIP 151: It's best if a colony doesn't produce its own queen

Since the bees know the mechanics of their lives far better than any human beekeeper, it should be logical for beekeepers to think that the bees can do a better job of removing and reinstalling queens than humans ever could. The thought is to 'Let them handle it themselves'. Though bees have millions of years of instinct invested in growing and installing queens naturally, they do not always do a great job of it. Maybe if humans had not delved into the genetics of bees so often the bees would be more proficient at this natural queen production process. The biggest problem for the beekeeper is that he or she never knows exactly when the supersedure process will start. Once it does start, about 50 days pass before that colony once again is producing adult worker bees. Natural requeening does not always provide a perfect solution for a queenless colony.

BEE BEHAVIOUR

TIP 152: *Become familiar with the bees' dance language*

Inside the dark hive, bees perform a complex series of movements (dances) that apparently give information about floral sources, potential nest sites and water source locations to other foragers. Inside information directs the bees when they are outside. Formal information relating to distance, direction, taste, odour and abundance seems to be transferred from a successful forager to other foragers. This complex communication system makes honeybees exceptional foragers and able to quickly exploit food resources within their environment. With a basic understanding of the dance biology, a beekeeper can watch dancing bees in an observation hive and get a general idea of where the bees are foraging.

TIP 153: *Know that some behaviours are not understood*

As is the case with many other poorly understood colony behaviours, bee washboarding behaviour continues to defy a clear explanation. Rarely seen in Europe, but popular in the US, bees form into distinct rows, literally elbow to elbow, and, with front legs lifted, they appear to lightly scrub the surface of the hive. Back and forth, back and forth they go for days on end, all in neat, nearly synchronous movements, very similar to the 'Mexican wave'. It has been postulated that the bees are laying down propolis layers, but studies looking at the surface afterwards do not see a notable propolis layer there. This propolis-masking behaviour at the entrance is common in tropical honeybees, and propolis use around the entrance in temperate colonies is common, but propolis application does not seem to be the cause in this case. The bees perform this activity in plain sight, but beekeepers are not privy to why the activity is being performed.

TIP 154: *Learn to read queen cells and their production*

❋ Actual queen cells in a colony always mean something to both the bees and to the beekeeper. Bees will only produce queen cells for ❶ swarming behaviour, ❷ queen replacement behaviour and ❸ emergency queen replacement behaviour. Any one of those behaviours is important to the observant beekeeper. A swarm will mean lost colony resources. Queen replacement behaviour (supersedure) may occur at a bad time and emergency queen replacement means that the queen died abruptly.

Swarm cells are generally the most numerous queen cells produced, with as many as 15–20 cells being produced. Swarm cells can be found nearly anywhere on brood combs. Supersedure and emergency cells are fewer in number; they are generally found well within the brood nest confines and not along the edges. Supersedure cells are felt by many to be the best-quality cells, while emergency cells are possibly inferior. In some instances during emergency replacement, the nurse bees do not have the necessary time and resources to produce excellent queens.

TIP 155: *Understand the queen's biology and mating behaviour*

❋ Even if a beekeeper does not use his hives to produce queens, all beekeepers need a basic understanding of the queen. A queen larva needs abundant food and nurse bee attention. The colony must be healthy and robust in order to produce good, healthy queens. The queen must have the right kind of weather to take mating flights and on those flights she must encounter sexually mature, robust drones with which to mate. This specialised queen behaviour is critical to the colony's survival.

TIP 156: *Don't believe that drones chase an unmated queen from her colony*

Beekeepers are frequently admonished to encourage healthy drone production within the bee nest. It is felt that properly mated queens cannot exist without proper drone numbers. That notion is correct, but what is not exactly correct is that the drones in a specific colony are not necessarily being produced to mate with queens from that colony. Drones are primarily for spreading the colony's genetics to other colonies. When the queen mates with other drones, she brings new genes to her colony. There is a possibility that in the chaos of the mating procedure, a queen will encounter a drone from her home colony. However, if all is going well, surrounding colonies will contribute drones to the community drone population. Mating may take place with up to 20 drones that are genetically unrelated to the queen. Drones from the queen's own colony do not pursue her as she leaves her colony for nuptial flights.

TIP 157: *Drone flight activity is an indicator of colony health*

Only a healthy colony will produce vibrant, healthy drones. On average, a strong natural colony of about 50,000 bees will have about 600 drones during the main season, so about 1–2% of the colony's bees will be male bees. The beekeeper, by watching for flight activity, should expect to see large, healthy drones leaving the hive in expected numbers. Significant numbers of undersized drones can mean a problem with the colony, while no drones of any size can signal that something is amiss within the colony. When watching drone activity, there should be a sense of balance. Many workers should be coming and going with pollen loads and the occasional drone should be both leaving and returning. Too many drones or not enough drones indicate a possible internal problem. Occasional healthy drones should be the norm.

TIP 158: *Watch the landing board for signs of activity*

Observing the landing board activity of a bee colony is enjoyable to any beekeeper. During the productive season, the entrance activity is intense and busy. Bee sorties are constant and entrance monitors are on guard for neighbouring bees or other pests trying to gain access. Worn out and diseased bees will unceremoniously leave the colony and drop on the compost heap in front of the colony. The beekeeper can get an easy appraisal of the dark hive's situation just by watching the entrance.

A vast numbers of bees (thousands) flying could mean that swarm preparations are underway or absconding is occurring. Just a few bees flying on a nice warm day could indicate serious problems inside. Dead and dying bees in front of the hive in excessive numbers are not a good indicator. Mounds of dead bees out front could indicate a recent pesticide hit. Bees fighting with each other on the landing board tell the beekeeper there is an ongoing nectar dearth, and hive openings should be limited to prevent outright robbing.

TIP 159: *Appreciate the risk a foraging bee takes with flight*

A bee flies about 12–15mph (19–24kmh) in normal foraging runs. She can fly up to 20mph (32kmh) if not loaded and the weather is suitable. A foraging bee faces real risks each time she leaves the hive. Spiders and other predators are waiting along the way. Wind gusts and rain can cause great difficulty and, though it is not known how often it happens, a bee can simply become lost. Foraging is dangerous work.

TIP 160: *Respect the hard work that house-cleaning bees do*

The amount of labour bees perform is nearly inconceivable to humans. Apparently bees do rest and seem to have periods of inactivity that can approach something similar to sleep, but in general bees are on the job for the 5–6 weeks that they live. Hygiene standards vary from colony to colony, but in general, the bee nest is a clean place. Workers are constantly polishing, building, cooling, heating, feeding and defending. While no part of a bee's short life is particularly easy, cleaning and maintaining is a major component of a healthy colony. Contaminants and trash that cannot be removed from the colony must either be coated in propolis or eaten and removed indirectly. For instance, bees crushed in a normal management event must be taken apart and removed from the colony piecemeal. Any liquid or body parts that cannot be removed must be consumed or coated in propolis. Keeping the hive clean is hard work and important to the health of the colony.

TIP 161: *Know the difference between 'scenting' and 'fanning'*

At the hive entrance, if a few bees can be seen furiously fanning their wings, they are normally either scenting bees or fanning bees. Bees can perform both scenting and fanning at the same time. Scenting bees will have their Nasanov gland at the tip of their abdomen exposed. The odour of this scent is akin to clean straw and serves as a chemical marker to bees that might otherwise be lost. While the gland can be used for different reasons, orientation is always at the core of its function. Alternatively, a fanning bee is simply passing air into the colony for a cooling response. It is not uncommon for scenting bees to also be fanning bees. When a beekeeper has just hived a swarm or moved a colony, the presence of scenting bees indicates acceptance of their new home. Large numbers of fanning bees alert the beekeeper to a colony that may be overheating.

KNOW HOW TO AVOID STINGS

TIP 162: *Respect aggressive, tempered bees*

During summer months, the nectar flow has generally subsided and thousands of bees are at the hive – essentially unemployed. It probably is not that the bees are testy simply because they are hot, but rather that there are just so many of them with little to do other than guard the hive. These bees have short tempers and will attack in larger numbers than usual. Regardless of the reasons, expect bees during hot summer months to be feisty and defensive. If possible, leave them alone.

TIP 163: *Use a smoker to disrupt defence within the dark hive*

The traditional hive smoker (see chapter 2) is a smelly but necessary piece of beekeeping equipment. The exact response of bees to smoke is still not understood. Smoke seems to mask the chemical communication system within the dark hive, thereby nullifying the guard bees' coordinated defensive responses. If smoke is accomplishing nothing more than simply masking the communication system, then why do other concoctions such as peppermint oil sprays or other strong odours not have a similar response? For whatever reason(s), smoke works. So, use smoke as sparingly as possible, but use it. Populous colonies during a nectar dearth are nearly impossible to manage without smoke.

TIP 164: *Remember, the occasional surprise sting will happen*

Stings and stinging are a normal component of keeping honeybees, but in no instance should excessive stinging (about a dozen stings) be considered normal. Experienced beekeepers are better at perceiving a colony's anxiety level, but predicting the occasional solitary sting is difficult. Such 'undeserved' events frequently happen to individuals who are standing farther away from the colony. Why a single bee decides to sting away from the hive is not understood or expected, but this unexpected sting is certainly an indication of the colony's current demeanour. Be careful.

TIP 165: *If stung, remove the detached stinger*

The traditional recommendation when stung is to quickly remove the venom sac that will remain after a honeybee sting. This is not bad advice, but it may not be absolutely correct advice for every sting. The stinger should be scraped away rather than pinched away. It is thought that the pinching action will force all the venom content into the wound, making the sting reaction worse. The problem is that all stings are not the same. Simply stated, some stings are more painful than others. While venom is of similar composition, the age of the bee, the amount of venom that has been produced, the temperature of the day and the location of the sting on the human body are important when evaluating the pain of a sting.

TIP 166: *Try not to fret about stinging away from the colony*

A honeybee can administer a sting at any point in her life, but she is much more inclined to sting if her colony is threatened. Bees in fresh swarms are usually very gentle, as are foragers on flowers. However, if a forager is captured and handled, obviously she will sting. This defensive sting is not for her survival, but for the safety of future bees.

SWARM FLIGHTS

TIP 167: *Develop a sense of swarm biology and how to deal with a swarm*

🐝 As recently as the mid-1980s, beekeepers could easily acquire new bees through the collection of swarms. Today, swarms are coveted and are rarer than in decades past. Today's beekeeper still needs to be prepared for the occasional call. Under good conditions, a swarm will only hang around at the original site for a few hours.

 A swarm that has only recently landed is normally a gentle bunch of bees. At that moment, they are homeless and will frequently accept any dark domicile that is offered to them. Unless vacuum devices are available, a beekeeper will probably go to the swarm site, move the swarm to the beekeeping equipment, and then return later that evening to get the bees after they have all settled. If the swarm is moved before all the bees are in the new equipment, a small cluster can remain for several days and become testy to anyone nearby. A proficient beekeeper should know how to predict and manage a swarm effectively.

TIP 168: *Explore the behaviour of play flights*

On pleasant late spring days, 'play flights' or 'orientation flights' may be taken by young bees in order to learn the lay of the land. The activity can become so great as to resemble a swarm issuing from the colony. The bees are generally hovering a bit, and darting rather erratically, but for the most part, appearing to go nowhere. This is one of those areas where apicultural science thins out. In fact, while orientation flights do appear necessary and while it is not doubted that young bees do take them, all such instances of large numbers of hovering bees have not proven to be young bees. Recent studies have shown that some of these episodes involve older bees that are undertaking this hovering procedure. Some professionals are now wondering if these older bees are preparing to accompany a new queen on her nuptial flight. If this proves correct, this behaviour will add even more complexity to the queen mating process.

TIP 169: *Bees may be attracted to neighbouring water sites*

Foraging bees will not respect human property boundaries. When searching for water, bees will seek a dependable source that is visible and within flying distance. Too often, such a site is on neighbouring property at a swimming pool or at a bird watering site. One bee, or two or even three, is not an issue. During the hot summer months, strong colonies may send hundreds of water foragers to dependable sites. Suburban beekeepers must include a clean, dependable water supply for their bees. Unfortunately, even then, the beekeeper can expect bees to visit neighbouring water sites. This behaviour will happen only during very hot months, but it can be a frustrating time for both the beekeeper and neighbours. When bees are seen at a watering hole near an apiary, certainly some of these bees will be from the beekeeper's hives, but not all. The beekeeper can only make sure that an abundant, dependable water supply is provided on their property.

TIP 170: *So, try to prevent your bees from visiting swimming pools or ponds*

Numerous tips have related to water supplies for your colonies. The reason for this is obvious – large numbers of honeybees around swimming pools or landscape ponds are a nuisance that can supersede their value as a pollinator to the pool owner. The chlorinated water has a strong odour cue and the vivid blue water is well within the bees' visual colour range.

Additionally, pool water has a distinct taste so other hive bees can be recruited to the source. Also, pools and ponds are dependable water sources. This problem of bees visiting the pool is a difficult one. Try not to put your bees near a pool and always provide your own dependable water supply.

TIP 171: *Fences will force departing bees to fly high*

🐝 Fences around your apiary are good for several reasons but one of the primary reasons is that it forces the foragers to gain altitude quickly and fly over the heads of neighbours. Shrubbery and trees will also achieve the same objective. Without fencing, returning bees will come in a long, low glide path and can potentially crash into a bystander. In most cases, barriers will prevent such an occurrence.

TIP 172: *Learn the difference between orienting flights and a swarm departing*

🐝 When a populous colony is exhibiting play flight behaviour, it can resemble a swarm departing. The two behaviours are considerably different. After an hour or so, the play flight behaviour quietly comes to an end, but if the behaviour is actually swarming behaviour, things are just beginning. Many times a swarm will initially camp near the original site for a few hours to a few days. Unfortunately, some swarms already have a new home site selected and leave the original colony and fly directly to the new site, which can be miles away. Obviously, these bees are lost to the keeper. Another possibility is that the swarm is preparing to leave, but is not yet able to pull it off. This pre-swarm behaviour can give the beekeeper a small window of opportunity to divide the colony before the primary swarm departs. Swarming is the natural behaviour of a healthy colony, but few beekeepers feel good about their bees flying away.

TIP 173: *Provide supplemental feed sources outside the hive*

🐝 Open feeding is the technique of pouring sugar syrup in large tubs or putting pollen substitute in open communal containers. In the past, university projects have open-fed pollen supplement dust with good results. Since open feeding is easier to implement than inside feeding, some commercial beekeepers with large numbers of colonies use this outdoor feeding procedure to cut time and equipment requirements.

There are no individual feeders within each hive, though there are significant concerns about open feeding. Although fast and easy, open feeding is a viable way to transmit diseases and can readily encourage robbing. No doubt, open feeding is better than not feeding at all. Try to use open feeding only during early to mid-spring seasons of the year and do not use it during times of nectar dearth when bees are defensive and prone to robbing. Robbing is a natural aggressive behaviour that is nothing short of ruthless for the colony being robbed. Don't encourage it.

TIP 174: *Realise that you have little control over outside bees*

🐝 Smoke only subdues bees that are inside the hive. For smoke to be effective on outside bees, the entire area would have to be fogged with dense smoke. When trying to discourage robbing behaviour or when trying to help another beekeeper who is under siege by irate bees, smoking bees in flight will have minimal effect on them. Bees will move away from the smoke, but they will immediately return as the smoke clears. Try not to get to the point where outside smoke is needed.

TIP 175: Don't automatically assume a foraging bee is yours

✿ While most people understand the value of bees for food production, they do not necessarily want them nearby. Community beekeepers often become the perceived source for all the bees in the neighbourhood. Usually, this reputation is a good one, but there are times when a nearby person does not want bees buzzing too close. In typical conditions, all bees seek unexploited blossoms – whether or not your bees are in the community.

Simply moving your colonies will not guarantee to a concerned neighbour that foraging bees will no longer visit his or her flowering plants. It is nearly impossible to simply look at a foraging bee and tell if it is from your colony. Determining if an individual bee is from your colony requires DNA testing, and such a test would only be performed (if ever) in legal situations. Don't automatically suspect that a specific foraging bee is yours. All bees look a lot alike.

TIP 176: Help returning bees from drifting to the wrong colony

✿ Most apiaries are not set up in a practical way for bees. Beehives, painted similar colours and aligned in bullet-straight rows, provide a confusing community for bees. Bees clearly see colours except for reds. Bees can clearly see ultraviolet, but that is of little help to humans. Painting the hives distinctive colours and putting colonies farther apart and not in rows will help bees find their proper homes. Otherwise, experienced beekeepers expect colonies located on the row ends to acquire increased populations from the centre colonies. The hives on the row ends are easiest to find. Periodically rotating or moving brood frames will help to adjust populations.

TIP 177: *Know the signs of laying workers*

🐝 Beekeeping around the world is currently enjoying a dramatic increase in popularity. This interest comes at a time when it is a challenge to keep bees healthy and alive. Since Varroa mites prefer male bees, drone populations have taken a hard hit, too. Honeybee queens are in demand and drone populations are stressed. Laying workers occur when a colony becomes queenless and stays queenless long enough for worker ovaries to enlarge. The time for this morphological change varies from a few days to a couple of weeks. The most obvious symptom is multiple eggs per cell found in the brood nest. The bees become noisy and the colony weakens with too many drones. This colony is beyond hope. The only solution is to combine it with another, stronger, colony.

TIP 178: *Discourage robbing behaviour*

🐝 Weak colonies, especially laying worker colonies, are prime targets for being robbed by stronger colonies. Don't leave combs outside, and, if possible, do not allow weak colonies to be robbed. Having robbers become familiar with the process will only embolden them to attack larger, more promising colonies. Do whatever you can to prevent robbing behaviour.

TIP 179: *Watch for dead pupae on the landing board*

Seeing dead drone pupae on the landing board is not a good sign, but the clearest indicator that something is seriously amiss within a colony is seeing dead worker pupae on the landing board. Starvation or extreme Varroa predation can be the common causes for pupae dying and being removed, as is with chalk-brood. In the scheme of dire events happening, drone pupae will be sacrificed before worker pupae. Since the pupal stage is the resting stage, all feeding has ended and the colony food investment has passed. If a colony is under great stress, it will eat the larval stage first, for that is the stage that will still require colony resources. Eggs and pupae will be initially spared, but if starvation conditions continue, the eggs and pupae will then be consumed. A colony in this state is in critical trouble.

TIP 180: *Know that crawling bees at a hive entrance is not a good sign*

There is not a single instance where significant numbers of crawling bees at the front of a colony is a good sign. The severity of the situation depends on how many crawlers there are. There are multiple reasons for bees crawling away from the hive. Parasitic mites or Trachea mites commonly known as Acarine, Nosema infections, viral infections vectored by Varroa mites or subtle exposure to a pesticide, are common reason why bees become too weak to fly. Aged or weakened bees seem to have some kind of suicide inclination that prompts them to leave the colony, thereby causing less work and pathogen exposure for the surviving colony members. Crawling bees means the beekeeper needs to do something – soon.

TIP 181: *Monitor comb building on new foundation*

Increasingly, beekeepers are accepting the convenience of beeswax-coated plastic foundation inserts. The problem is that the bees are less than enamoured with, what to them is, a foreign product. Always use 10 frames when providing bees with new plastic foundation. If given even the smallest amount of extra space on this new foundation, bees will sometimes begin constructing distorted and disfigured combs on these inserts. Yes, the beekeeper can destroy the comb and force the bees to do it again, but unless all of the previous comb is scraped away, the bees will do it wrong again. Put the colony on 10 frames, and if the flow is weak, provide supplemental sugar syrup to encourage the bees to build good combs. After plastic combs are used correctly, all subsequent use will be normal.

TIP 182: *Watch for uncapped honey*

If a nectar flow is abruptly terminated, house bees may not have the wax resources to finish capping the honey crop. The uncapped honey may very well be at the correct moisture level (18.6% water), but the cappings are not in place. The uncapped honey has a tendency to reabsorb airborne water which, in time, can ferment. The simplest solution is to extract the uncapped honey and blend it with honey that has been capped and at a moisture level lower than 18.6% moisture. Not having this low-moisture honey for blending, the only other solution is to leave it with the bees for their use. Possibly putting on supplemental feed will initiate wax production that can be used for the cappings. This procedure is only practical for a few frames.

COLONY
MANAGEMENT

Honeybees have become ever more dependent on their keepers during the last two decades. Diseases, pests, insecticides and changing ecosystems have come together to stress the honeybee population around the world. More than ever, sound manipulation and management schemes are essential for keeping our bees healthy and productive. You need to know your stuff!

FEEDING AND WATERING

TIP 183: *Follow through with colony feeding*

Feeding colonies supplemental carbohydrate is a serious undertaking. If the colony is starving, in late winter or early spring, this colony will require a consistent supplemental carbohydrate supply. Erratic and inconsistent feeding will only be a waste of time and money. Survival rations should contain as much sugar as possible (at least 2 parts sugar to 1 part water). Stimulative rations are fed to kick start the spring season and get the bees to begin producing brood. This syrup can be thinner (1 part sugar to 1 part water). Feeding for survival is tricky. The bees might have to go back into a cluster and be away from the sugar feed. Dealing with the extra water that the syrup contains is also a problem for the bees. Fondants and dry sugar formulations are possibilities. Be serious about feeding needy bees.

TIP 184: *Consider using top feeders for sugar syrup*

Division board feeders and entrance feeders are probably the least useful for supplying the supplemental carbohydrate needs of wintering bee colonies. Fondant on the top bars works OK. Consuming dry sugar requires that the bees find water, so make sure a water source is available nearby. Also, there is no reason to put on more feed than the bees can consume in a few days. As the weather warms up and bees move about, top feeders, which are tray-like devices on the very top of the hive, are a good source. With today's busy schedule and to save your bees, create a few pinholes in a bag of sugar, hold the bag under water for a few minutes and place on top of the crown board. However, individual bees must learn how to use these feeders, and, as they age, wooden top feeders may begin to leak.

TIP 185: *Feed colonies a protein substitute*

⚜ Feeding a protein supplement is more expensive than carbohydrates and the bees are sometimes not inclined to eat the supplement. Most come in a solid powder or cake form – there are a few products that are liquid, but these are not always readily available. If natural pollen is added to the supplement, the consumption rate increases considerably. (Any pollen added must be disease-free.) As is so often the case in beekeeping, if the feedings are successful and the colony builds up, then the beekeeper will have to begin watching for swarming issues (see tips 167–182).

TIP 186: *Be aware that supplement consumption will vary*

⚜ Some colonies are simply faster learners than others. Maybe some colonies are more desperate, but, for whatever reason, some colonies will take supplemental foodstuffs faster and better than others. There's not much the beekeeper can do other than offer the products to the wintering colony. Different styles of feeders require individual bees to learn the mechanics of the feeder. Sometimes feeding works and sometimes it doesn't.

TIP 187: *A dependable water source is critical*

⚜ The colony's primary uses of water are to dilute honey and to cool the hive during hot seasons. While some metabolic water is available to the colony, external water sources are a necessity. Dependable water sources that are not allowed to run dry should be cleaned about once per week. Strangely, a significant number of bees drown in the water source. Floats or stones are commonly used to give the occasional clumsy bee an exit. Bees will use water from nearly any source, including dew and standing puddles. Water-foraging bees do not always select the most pristine water source.

WHERE TO PLACE A COLONY

TIP 188: *Think carefully about where to locate your apiaries*

In general, beekeepers put their colonies where they can – not where they should. Many recommendations exist in both new and old texts. Some say to face colonies to the south to avoid winter wind blasts. Others recommend facing colonies east to encourage the bees to awaken early to get an early start on bee work and foraging.

Without any help from you, these colonies are amazingly aware of the time of the season and, indeed, even the time of the day. They instinctively know when they should initiate brood rearing and food foraging. Our beekeeper efforts of supplying 'stimulus' sugar feedings probably don't help, but neither do they hurt. If a colony does not seem to be aware of the season and what they should be doing, the traditional recommendation of queen replacement will become important.

TIP 189: *Put colonies in the sun*

As a fully populated colony becomes crowded during summer months, in order to stabilise the internal temperature at about 36°C (96°F) most of the non-nurse bees simply move out of the hive thereby helping to reduce internal colony temperature. No doubt water foragers will still be required, but a mass of bees on the front of the colony will be the result. The hot cluster may hang on the front of the colony and show signs of overheating, but if they have easy access to water they are OK. Essentially, such hot bees are, 'Sitting on the front porch'. They can shed water, but bees exposed on the front of the colony would be particularly vulnerable. If possible, do put colonies in the sun but add more space at the hive front.

118

TIP 190: *Make sure your hive equipment is up to standard*

🐝 Certain pieces of hive equipment are useful, but not commonly used. Slatted racks and dummy boards are examples of old equipment pieces that are not widely used today. A slatted rack is a wood frame having narrow wooden slats running from side to side with 1 cm (³⁄₈ in) openings between each slat. It is positioned on the bottom board and provides the colony with more clustering space. A dummy board is simply a temporary wall exactly the size of a frame – for instance, it allows the beekeeper to partition a 10-frame colony into a five-frame colony. Alternatively, new expanded polystyrene hive equipment has interesting insulating characteristics. Don't be afraid to look at new and uncommon equipment when you are thinking about where to place your colony, for both aesthetic and practical reasons. (See chapter 2 for more on basic beekeeping equipment.)

TIP 191: *Find solid ground for your beehives and vehicles*

🐝 Accessibility to the apiary is a key feature of a good location. Ideally, both pathway and apiary surfaces should be solid. Soggy ground makes carrying heavy equipment difficult and can allow the stands beneath the hives to sink as ever-increasing weight is gained during the season. An unstable driveway will be difficult to travel during rainy seasons. Since honey is heavy, a vehicle can occasionally stall on muddy driveways. The common response to this situation is to visit the apiary only when the ground is dry. That requirement really limits the beekeeper in properly managing the colonies and bringing in equipment. A firm footing for the beekeeper, the beekeeper's vehicle, and for the bee colonies is a good feature of a productive apiary.

TIP 192: *Don't worry too much about hive orientation*

Think about it for just a minute. The entrance to the typical colony is approximately 1cm (³⁄₈in) to 2cm (³⁄₄in) wide. There is no way that light from that narrow entrance is going to penetrate inside a tall hive. Simply stated, the bees know what time of year it is. Again, there is no harm done if a beekeeper is providing sugar syrup and spring supplemental protein, but the bees know the season. They are not waiting to be awakened. So, should you face your colonies to the east? Sure, face away. Again, no harm done, but this directional mandate, which is presented in so many texts, is not critical. The air flow in the apiary, the general morning sun (are they in the shade or are they in full sunlight?), the general health of the colony and the queen's vigour are all much more important factors. Orientation with the sun does not make or break a apiary location.

TIP 193: *Don't always keep hives in straight lines*

The traditional, photogenic apiary is commonly made from nicely painted white hives in perfectly straight rows. While this is visually pleasing and makes hive and bee husbandry management easier, it is not the best layout for bees. The hives all look alike and finding the right hive is difficult for returning forger bees. Hives on either end of the row tend to collect population while colonies in the middle of the row tend to lose population. There are many recommendations for facing the hive or positioning the hive, but hives that are positioned fairly randomly with entrances in different directions are easier for bees to find.

TIP 194: Expect to have to deal with gates

Most migratory beekeepers would place their hives in open fields or on farmers land with granted permission. Gates are a common barrier which many beekeepers must deal with. Some gates are good while some are annoying. The presence or absence of a gate is rarely the death knell for an apiary location. Gates commonly require keys (which are occasionally forgotten), and they require opening and closing. Openings that require turning off narrow roadways make it difficult to align the vehicle well enough to pass through. But all the negative features discussed above also protect the apiary from thievery and vandalism. Good gates can readily be a positive feature of a productive apiary.

TIP 195: Remember, farm animals and bees don't always mix

Domesticated farm animals are no strangers to insects and pests. This includes bees. Conversely, bees are no stranger to animals and other insect pests. Normally, bees and farm animals coexist nicely without issue. However, a bee colony that is being pestered will sometimes attack any nearby animal regardless of fault or blame. Obviously, the animal will move away if bee stings are severe enough. But if the animals are penned or tethered, a stinging incident can occur. Large animals like horses or cows have a much better chance of withstanding the stinging attack, but small animals, such as chickens in a coop, are sometimes in harm's way. If the beehive is away from such interactions or if the animals are free to move away from the colony, there is rarely a problem.

TIP 196: Think of your neighbours

🐝 Faecal spotting, swarm settling, water foragers at bird baths and the occasional sting can really antagonise a beekeeper's neighbours. If it seems practical, tell your neighbours what you are doing. If what you are doing is legal by current regulation, hiding your colonies or using a bit of camouflage may be appropriate to help with good neighbour relationships. An 'in your face' attitude rarely works out.

TIP 197: In towns or cities, put your bees on balconies or roofs

🐝 It may not be the best producing site, but an apiary can be located very high from the ground. Moving the colony to or from the high-rise apiary is generally more demanding than locating a colony in a normal apiary. The beehive equipment must be totally sound, without holes and leaks. The entrance and the top should be screened tightly, preferably with a migratory screen or heavy open mesh linen cloth. Ideally, the cart used to move the colony should have pneumatic tyres to soften bumps along the way. Obviously, elevators must be used only when human traffic is lowest – probably very early in the morning. Building regulations, council regulations and neighbours must be considered.

TIP 198: Paint your hives any colour but black

🐝 Pure white is the most common colour for painting beehives because it was believed to keep the hive cooler. In reality, any colour is seemingly accepted by the bees except pure black, due to heat accumulation. Bees have a general dislike for black-coloured protective clothing (hats, gloves) and for black watch bands – a common reason given that black is the colour of honey badgers and other pest animals. Any type of paint or stain can be used so long as the finish is allowed to fully cure – not just dry – before use.

TIP 199: *Tackle cleansing flight mess as much as is possible*

For obvious hygienic reason, bees defecate some distance from the hive. The excrement appears as small yellowish splotches that are no more than 6mm (¼ in) diameter. Presently, there is no known way to manipulate the corridors that bees use when taking cleansing flights. Sometimes cars or neighbouring houses are the unintended targets for these aerial discharges. Unfortunately, these fecal splotches are acidic and can etch paint finishes on vehicles. If possible, don't park vehicles near the hives.

TIP 200: *Avoid pesticides, but know that they are everywhere*

Apiaries in any location are exposed to a greater or lesser degree to pesticides. Pesticides cannot be totally avoided in most places. Even so, the beekeeper should check the surrounding areas to determine how much insecticide is commonly used. Older chemical classes of insecticides left obvious piles of dead bees in front of the hive. Modern insecticides, though safer and more environmentally friendly, are more obscure in their effects on bee colonies. In some cases, it seems that the bees' orientation abilities are affected, and they are unable to find their way back to the hive. In this case, the only real symptom is a declining bee population within the hive.

Within beekeeping, pesticides and their use – both inside and outside the hive – are an emotionally charged issue. No doubt, discussion and research will continue in the future. Until a resolution is developed, the beekeeper should strive to protect the colony from pesticides.

MOVING HIVES

TIP 201: *Know how to relocate hives*

🐝 A feature of modern beekeeping is the ability to move a colony great distances when necessary. Commercial pollinators do this all the time. However, even hobby beekeepers must move the occasional colony from one apiary to another. The usual procedures necessary for relocating colonies are well known, but these procedures are not precise. Every bee colony move event has its own unique factors: How heavy are the colonies? How far is the trip to the new location? How sound is the hive equipment? What will the seasonal temperature be during the move? The beekeeper must factor all these variables into the colony move plan. But a few features are firm: Screen the entrance and the top for ventilation and keep the colonies cool – even if it means pouring a few cups of water into the screened hive. Don't let the colonies overheat or suffocate.

TIP 202: *Ask a bee friend to help you with hive moves*

🐝 The weight and size of beehive equipment will nearly always require two people to be involved in a relocation procedure. Even if the beekeeper is strong enough to handle a full hive with two deep supers weighing 68–90kg (150–200lb), it is still too bulky for one person to handle. A multitude of carts and hand trucks are available, but a second person is very useful for stabilising the load as it shakes along the way. Hive straps are commonly used to secure the equipment for the move but, even if tightly secured, the equipment can sometimes twist or move a bit, allowing frustrated bees to escape. It will be impossible to quickly get the freed bees back into the equipment. Yes, one person with the proper equipment could make the move, but a beekeeping friend is a real asset for a beehive move.

TIP 203: *Make a detailed plan before moving a colony*

⚜ Since the bees are all inside the hive at night, evening is a common time to move bee colonies. This night work is always an adventure. Beekeepers must wear full gear, so the veil makes vision even more obscure. Smokers become nearly invisible and a hive tool is nearly always lost in the darkness. But most challenges are made easier if the weather is cool. The bees stay calmer, and the beekeeper is not as hot. To a degree, a daytime move is possible, too.

The free-flying colony is smoked a bit and screens are put on the entrance and on the hive top. Then the confined colony – with bees flying all about – is put in the vehicle. A small 'bait colony' must be left to gather all the field bees that are out flying. If the bait colony is queen-right, the queen should be caged for her safety. About one 10-frame trap colony should be left for about every 10 colonies moved. This area will have a high population of testy bees, so this procedure should not be done in populated areas. At the beekeeper's convenience, the single-story trap colony can be easily and quickly moved at night. The bait colony's fate is up to the beekeeper.

TIP 204: *Have multiple flashlights on hand*

⚜ Good torches are fundamental to the success of a night move. They need to be heavy-duty and dependable. If the weather is warm, defender bees will be drawn to the light. Red filters on the torch may help some with this characteristic. Having multiple torches for the move is essential.

TIP 205: Do not depend on propolis to hold the hive together

🐝 It is very tempting to try to move a colony that seems soundly stuck together with propolis. Since the propolis joints are brittle, trying this trick is nearly always a bad idea. Even though it is called 'bee glue', propolis is little more than a natural filler compound and is not an adhesive. Hive straps or nylon bands are common strapping materials used to secure beehive components. Commercial beekeepers sometimes use metal strapping. In years past, beekeepers used large staples that are still available from bee equipment suppliers. Wooden slats or metal brick ties have also been used. Don't depend on propolis for even a short move.

TIP 206: When moving bees, wear your best bee suit

🐝 On either a night or daytime move, if the weather is warm, this is a perfect time to have a highly protective bee suit. Bees are highly defensive of their nest and all the jostling and tumbling getting the hive ready for loading really alerts the bees to the upcoming event. In this regard, smoke is a great but incomplete help. Smoke will cause bees to delay a bit in flying to defend the hive, but it will not completely stop them. Bees in the air cannot be smoked. So use smoke, but your primary defence against angry bees will be your bee suit.

TIP 207: Expect relocated bees to be unhappy bees for a while

✺ Due to the concentration of both carbohydrate and protein storage within the nest, bees are accustomed to having to defend themselves. When a bee colony is moved and bumped, guard bees throughout the colony are alerted. In general, the entire colony becomes aggravated. The bees left behind are frustrated and may sting. The colony are certainly aware of the moving event. Bees will jam behind the entrance screen attempting to find an exit. When the entrance screen is removed, the beekeeper should not be surprised to find that the bees are upset, and wearing a bee suit is important. It may take as long as three days for the bees to settle down.

TIP 208: Secure hive covers before moving hives

✺ Beekeepers must be aware that a move on an open trailer exposes the equipment to 55–65mph (88–104kmh) wind speed. If outer covers are not restrained by strapping devices, they are likely to go flying off the colony. Migratory beekeepers frequently use migratory outer covers that are little more than reinforced flat boards large enough to exactly cover the hive top. These flat board covers can be lightly nailed to the topmost deep, securely attaching them to the beehive. Moving live bees is not a typical load.

TIP 209: Expect bees to find ways out of the closed hive

✺ It only takes the tiniest crack for the bees to trickle out of a confined hive. Frequently, as a heavy hive is pushed and shoved onto the open trailer – even when strapped – there is just enough twist for a small crack to open. Unless it is a large opening or unless beehives are inside the car, these escaping bees are not a real issue. Bees have an uncanny ability to find such small exits. Duct tape or any other heavy tape can give a short-term fix. Screen wire can be stapled on for a longer-term fix.

TIP 210: *Take steps to prevent overheating in confined colonies*

An important aspect of colony moves is the surprising amount of heat a confined colony can generate and how quickly the colony can generate that heat. Keep in mind that this same cluster can produce enough heat to keep itself alive on the coldest winter day and cool on the hottest summer day. Bees excitedly searching for an exit generate considerable heat levels. The lethal problem for the confined colony is that the heat generated cannot be dissipated. Worker bees heated to about 36.1–36.6°C (97–98°F) will regurgitate the contents of their crop to cool their head and thorax. Consequently, colonies that have died from heat exhaustion will have a sticky, wet look. Don't spare the water if a confined hive is at risk of overheating.

TIP 211: *When possible, move colonies in the winter*

During winter months, dormant colonies can be gently moved onto an open trailer and relocated. No doubt the vibrations of the trip are upsetting, but so long as the cluster is not broken, the bees will continue to generate heat and survive quite well. Wintering bees make no effort to heat the insides of the hive but rather only heat the insides of the cluster. But the weather needs to be cold not just cool. At around 0°C (32°F), the cluster in transit will begin to generate heat. In fact, a confined colony can die of heat exhaustion when the outside temperature is around 0–1.6°C (32–35°F). If you have any concerns, use screening as you would during warmer months.

TIP 212: *Know how to find the apiary location in the dark*

❀ Thing look much different at night, and if the new apiary was visited a few weeks earlier, plant growth may have dramatically changed the looks of things. Moonlight helps but obviously is not always available. In pollination moves, new temporary sites must frequently be found in the dark or during early morning hours. The need to work in haste with live, confined bees, means that it is not totally surprising for a beekeeper to occasionally overlook some of the hives. Consequently, a second trip is required to retrieve bees that are still at this temporary location. Both the grower and the beekeeper are inconvenienced. A GPS device is a requirement for beekeepers who provide pollination services.

TIP 213: *Get hives opened before sunrise*

❀ While confined and still on the trailer or in the vehicle, bees generate a low, resonating hum. That hum is oddly unnerving for it makes it clear that the bees are pent up and unable to care for this colony. As dawn approaches, the hum increases, and the colony becomes outright active. Then the heat issue discussed above becomes real. The goal should be to get the colonies out of the vehicle before sunrise. The hives are easier to open and bees are less defensive.

TIP 214: *When necessary, move hives within the apiary*

❀ During winter months, hives can easily be relocated within the apiary. Just gently move them as you wish. However, colony relocations during warm months must be done slowly, maybe 1.5–3m (5–10ft) at a time over several days. If hives are moved to temporary sites for pollination, they should be away for about a month or so before being brought back to the original apiary. Relocating colonies too fast will result in drifting and disoriented bees.

COLONY GROWTH

TIP 215: *Allow space for colony growth*

During the spring and summer, colonies undergo a dramatic population increase. The winter cluster is about one-third to one-half the size of the summer cluster. If additional space is not supplied, the bees build swarm cells and leave. If too much space is supplied and the nectar flow is puny, the bees will not use the space and may even chew the comb to use the wax at other places within the colony. When the combs are built on that chewed foundation, there will be a good deal of transitional and drone combs on that frame. If an error must be made, it should be too much space rather than crowding the colony living area.

TIP 216: *Clean old combs before returning to the bees*

Increasingly, the practical life of honeybee combs is in question. Just a few decades ago, it was thought that combs could be used almost indefinitely. That was before the chemical pesticide knowledge base grew to what it is today. Wax comb is a bit like a sponge. It absorbs many chemicals with which it comes into contact. Additionally, much of the chemical residue found in wax combs was put in that hive by the beekeeper to control mites (see chapter 7). It feels wasteful, but the combs should be replaced every few years – probably every 4–5 years. Some beekeepers are already dating or colour coding their frames in order to keep a record of the comb age. Beekeepers who are replacing combs do it on a sequential basis – a few at a time during the nectar flow. This process is a good idea, but it is yet another task that the dedicated beekeeper must do to keep healthy bees.

TIP 217: *If using plastic frames, hope for good nectar flow*

🐝 Bees are not overly eager to build combs on plastic foundation that is suspended in either wood or plastic frames. However, even with the bees' poor opinion of plastic, because no assembly is required, beekeepers are increasingly using plastic frames and foundation, especially black foundation as the colour is very helpful to spot small eggs. To properly assemble a wooden frame and install foundation takes an estimated 20–30 minutes per frame (see chapter 2). Plastic frames come ready to use. The best way to coerce the bees into using a foundation medium that they are not completely sold on is to put the frames on during a good flow. Be sure to use 10 frames until the combs are built. Watch the combs' development. Bees will sometimes try to build in the space between the combs.

TIP 218: *Try not to waste the brood in a cross-comb colony*

🐝 It happens. A comb or even several combs are left out of a colony and the nectar flow begins in earnest. By the time the beekeeper returns to the colony and catches the error, the space has been filled with freely drawn burr combs. Though the bees seem to be happy with the situation, the frameless comb configuration defies traditional management schemes. Yet, much of the time, the queen has begun a brood nest in this natural comb matrix. It is best to deal with it while the flow is still ongoing. Use smoke and cut the combs from the box. Perform the same procedure used to transfer bees from a dwelling. Cut the brood combs and tie them into empty frames. Place them inside the colony above a queen excluder. Remove them when they are free of brood.

TIP 219: *Opening a colony is always disturbing to the bees*

✺ Predatory mites (see chapter 7) are troublesome for honeybees and beekeepers must help to control them, but helping means that the beekeeper must stress the bees by opening the colony. Novice beekeepers must open the colony to learn the techniques of good bee husbandry, but experienced beekeepers should increasingly only perform necessary tasks. In order to reduce hive disruption, bunch necessary tasks together. Bee colonies must be opened to manage the bees, but the beekeeper should work quickly and exit as soon as possible.

TIP 220: *Don't use any more smoke than necessary*

✺ Using smoke only subdues bees that are in the colony, and its effects last for about 10 minutes. After being subjected to smoke, worker bees engorge on honey and are not as defensive. Each time a hive is smoked and opened, much of the foraging efficiency for the remainder of the day is disrupted. Use as little as possible.

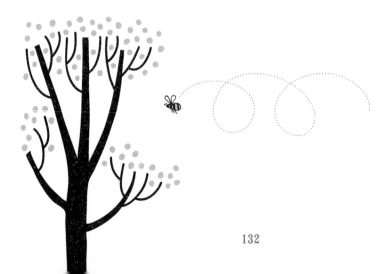

TIP 221: *Always try to be a patient beekeeper*

🐝 Beekeepers are much like gardeners. They must wait to see the results of their various efforts. If a new queen is introduced, after her cage is positioned in the colony, nearly a week must pass before she can be released into the colony. Then another week or so is required to be sure that she has been accepted and is laying eggs. Approximately 55 days must pass before the requeened colony is once again producing worker bees. Empty honey supers are put on the hive and the beekeeper must again wait to see if the bees will fill them or leave them empty. If swarm cells must be destroyed, the patient beekeeper must carefully scan the brood and then check it again a few days later. During winter months, the beekeeper waits to see if the bees will survive. A person with a steady, patient personality frequently makes a good beekeeper.

TIP 222: Don't spend time with weak colonies in late autumn

Some bee colonies just don't measure up. Maybe they were genetically inferior, or the nectar season was flat, or the bees were exposed to pesticides, or mites attacked. For whatever reason, a colony ends up being weak in the late autumn. Unless the colony is in a very warm climate, this colony has little chance to survive the upcoming winter. In warm climates, feeding these weak colonies might get them through the winter, but not without extra work from a concerned beekeeper. Commercial beekeepers cannot justify the time expended on unproductive colonies, but many times, hobby beekeepers take on these cases for enjoyment and exploring bee biology. Even so, much of the time, these colonies die in the winter.

TIP 223: Be realistic about saving a dying colony in winter

All beekeepers, especially newish beekeepers who have not experienced the death of a colony, feel some pain when they realise that a wintering colony is in trouble. The brutal fact is that little can be done to help the colony during a hard winter. Both the bees and the beekeeper should have done things differently late last autumn. Now, the colony must face its destiny.

TIP 224: *Know that a colony only stays good for 2–3 years*

Many beekeepers have favourite colonies that always seem to be the leading producer in the apiary. But the favourite colony will probably only hold that position for 2–3 years. Everything seems to change: the queen grows old, the mite load increases, bee combs continue to acquire chemical residues. There are many reasons for colony productivity to wax and wane. In general, a typical apiary of 10 or more colonies divides itself into thirds. One third is in great shape and currently needs nothing from the beekeeper. The middle third is struggling and can be helped by an experienced beekeeper, while the final third is simply in poor shape. The division may not be quite this bad, but over time, the productivity of colonies in an apiary comes and goes. Your best colony will not always be your best colony.

MANAGING a QUEEN

TIP 225: *A 'good' queen needs the support of a 'good' beekeeper*

No queen producer strives to produce poor queens. Many beekeepers stand ready to spend more money to be sure they install the best queen available in the colony. For most beekeepers, just having a 'good' queen should be enough. If a beekeeper is installing a great queen, she will need the support of a great beekeeper. Plenty of super storage spaces, a timely mite control programme and high-quality hive-management programmes that include a protein supplement and proper winter preparations are examples of tasks that the beekeeper needs to provide for his excellent queen. During most seasons, a good enough queen is good enough for most beekeepers.

TIP 226: *Remember, don't get too attached to a colony queen*

Queen management is complicated even for experienced beekeepers. Replacement queens are reasonably expensive and are not always readily available. The queen is either praised for the colony's success or blamed for its failings. Truly, the queen is the focal point of the colony (see tips 145–151). It is easy to see how a beekeeper could develop a fondness for a bug. The beekeepers know where the queen came from and how she was introduced. If she has been at the helm during a good honey crop or two, that provides even more reason for being attached to her. When she begins to fail, it is human nature to want to keep her as long as possible. The fact is that there is no practical use for an aging, failing queen. If she were left in the colony, the workers would do the deed for you. Queens that are OK can go for a while longer, but the beekeeper should know that an OK queen is not much of a colony monarch. Don't be afraid to replace her.

TIP 227: *Combine a queenless colony with a queen-right one*

By the time a new queen can be obtained for colonies that lose their queen late in the season, the colony will have begun a normal population decline. (The colony will be going into winter with a reduced population of workers that are older than they should be. Unwanted drones will be cast out.) In general, the colony has started making plans for surviving the winter. It is not the time to be requeening. Combine the colony with another queen-right colony and split the larger colony in the spring.

TIP 228: *You can never guarantee a queen's success*

There has never been a halcyon time in beekeeping when beekeepers were universally happy with their queen stock. The complaints voiced by beekeepers today are not new. Indeed, there is a reasonable chance that the fault is not even with the queens but rather with the drones available for mating with the queens. The diet may have been poor during her development stages. Maybe Varroa mites (see chapter 7) infected her with a virus? Should the beekeeper buy from the same producer or try another? The average beekeeper will probably never know the exact reason why a queen failed. For the foreseeable future, beekeepers continue to have to make decisions about a queen without all of the information needed to make those decisions

TIP 229: *Failing queens mean weak colonies*

Frugal beekeepers are sometimes tempted to keep failing queens going for longer. Today's honeybee queens only average about a single productive year. If the queen has already gone through a season and the colony has not excelled, something is not right. Either the queen is to blame or the season was not good. Either way, the queen's time has passed. The ideal situation is to introduce a new queen if it is not too late in the season (see tip 236 on introducing new queens). If very late in the autumn, uniting the colony will save the colony through the winter.

TIP 230: *Keep a healthy drone population*

Drones need more respect. They are specialised colony individuals whose major function is to mate with unmated queen bees. Only about 1–2% will be successful and the mating event will kill them. They will live about 30 days and as the winter season approaches any that are not successful in mating will be forced from the hive by intolerant worker bees. About 13–17% of a wild honeybee nest is dedicated to drones. At any given time, a healthy colony will keep about 400–600 drones. The colony's male bees are not specifically for copulating with the same colony's new queens. These drones are the colony's contribution to the honeybee gene pool. At this point in our bee knowledge, drones are specifically for mating purposes. A large population of undersized drones with a few workers is a strong indication of a queen problem that has allowed worker bees to produce eggs that only become drones. Big healthy drones are a sign of a productive colony.

TIP 231: *Address a laying worker situation right away*

🐝 Laying workers result when the queen is either defective or dead and replacement queen initiatives have failed. Under these queenless conditions and without suppressing queen pheromones, the ovaries of a few of the workers develop enough to produce a few eggs. Since the laying worker cannot mate, all of these eggs will develop into male bees. In short order, the colony will become imbalanced with undersized, unproductive drones. This colony is of little use and essentially has no future. Combine it with a viable colony or allow it to die out; then reuse the equipment.

TIP 232: *Laying workers can't be shaken out of the colony*

🐝 There is an old technique that is found in several older books that is still occasionally recommended by experienced beekeepers today. This procedure does not work. It recommends taking a laying worker colony 23–27m (25–30yd) away and shaking all the bees out. The equipment is then replaced on the original hive stand. In theory, the field bees will fly back, but the laying workers will be lost and remain at the shake site. This information is incorrect. Laying workers routinely fly outside the hive and may even forage a bit. There is no reason to think that they will become lost or shaken out. This whole process is pointless since the laying worker colony is small, weak, queenless and doomed anyway. Don't waste time carrying this hive around and shaking bees out of it.

GETTING LIVE BEES

Obtaining bees is an integral part of beekeeping. Several options are available to both the novice and the more experienced beekeeper. One way or another, the beekeeper must get bees – they can be purchased, or 'free' bees can be hived from swarms or removed from dwellings. Availability, seasonality and costs are variables that must be considered, so that you can find bees most suited to your requirements.

NUCLEUS BEES

TIP 233: *Package bees are not readily available in Europe*

🐝 In many countries, especially those of the US, Canada and Australia, some specialised beekeepers produce bees. The bees are shaken through funnels into packages with averaging 1.4kg (3lb) in weight. Included in the package is a suspended caged, mated queen and a can of feed for the journey. These artificial swarms can be transported in may ways, internal mail, carriage and even by air freight. Nowadays, most packages are produced by very large operations producing thousands of packages and tens of thousands of queens. However, in the UK and most of Europe where there are no packages of bees available, we rely on swarms, purchasing or making a nucleus of bees by splitting a strong colony or removing frames of bees from stronger colonies and adding a queen cell, or introducing a queen.

TIP 234: *If possible, combine your order with other beekeepers*

🐝 Most beekeepers in the UK purchase bees by way of a nucleus of bees, a frame of bees, or a new queen. The best way to get these is to combine your order with others and have them custom-delivered. For introducing queens, advanced beekeeping experience is needed. Bees (and queens) imported from inside the EU must carry a health certificate and other legislation must be adhered to (see tip 286). Most beekeepers purchase a nucleus of bees. These can become expensive from £150 upwards. Basically a nuc is a small hive containing five frames of bees. The nuc should have two frames of honey pollen and honey stores, and a young queen, plus bees of all ages. Bees (and queens) are normally ordered the in the previous autumn for delivery in spring or early summer of the next year. Beekeepers can also purchase a frame of bees which normally cost around £20.

TIP 235: *Release the nucleus of bees as soon as possible*

🐝 Any bees purchased must come from a qualified source free of disease and any other malady. The bees should be of good temper and a young queen of the first season. The equipment should be of good sound order, and the standard five frames of bees installed. The new nucleus should be placed away from strong colonies in case of robbing. The nuc need to be in a sunlit area out of winds and not placed under trees. The bees are likely to have been confined for some time so it is important to release them quickly.

TIP 236: *Introduce queens during the warmest part of the day*

🐝 According to DEFRA and the National Bee Unit (NBU), there are around 40,000 beekeepers in the UK, with over 200,000 colonies. About 300 are commercial managing around 40,000 colonies. There has been a huge increase in the importation of queen bees from the EU between companies (see tip 286 on importation). It is vital that on receiving a consignment of imported queen honeybees from an eligible country outside the EU, that you transfer the queens to new cages before introducing them to local colonies.

When introducing a new queen in a queen cage, the warmest part of the day is ideal. The colony or nucleus should be without a queen and check to see that there are no queen cells. Remove the attendants from the cage. Smoke the bees and lift the roof and crown board off the hive. Open the frames in the centre of the brood box. To hold the queen cage, remove the cork from the end of the cage containing the white sweet fondant. Pierce a fine hole in the fondant with a toothpick and place the queen cage sweet-side up in a vertical position between the two centre-most frames. Squeeze the frames together so the cage is held firmly in place. Replace the frames in the super. Replace the crown board and roof. Do not disturb the colony for a few days.

TIP 237: *Limit bee drifting by spacing the colonies apart*

❋ The 'drifting bee' phenomenon occurs even within established colonies. Regardless of the method of release that is used, some bees, new to the area, will become lost and drift to neighbouring colonies. Bee colonies – towards the ends of the row – will tend to be stronger and more highly populated colonies than colonies nearer the middle of the row. If practical, space colonies at least 3–6m (10–12ft), or more, apart. Additionally, not putting the colonies in perfectly straight rows is also helpful to the orienting bees (see tip 193). However, if the hives are set randomly, garden mowing and general hive maintenance may be more trying for the beekeeper. Loading bee colonies for movement to other apiary locations (see tips 201–214) may be extra work too. In general, staggered beehive locations are better for the bees, but not always better for the apiarist.

TIP 238: *Provide new bees a drawn comb frame or two*

❋ New beekeepers are frequently required to acquire or collect a swarm of bees. Swarms of bees require comb foundation. Approximately 21½cm (8½in) of honey must be consumed by bees to produce 2½cm (1in) of beeswax. If it is used to build honeycombs, that 0.45kg (1lb) of beeswax can store around 10kg (22lb) of honey. Clearly, giving the new colony some drawn wax combs is a help to the young colony. It provides the bees with a bit of space where incoming nectar can be processed and stored and the queen is able to begin brood production earlier.

Food storage and brood production are important in stabilising the young colony, thereby helping bees fully accept a queen that is initially quite foreign to them. Any comb given should be disease-free and probably only 2–3 years old. Increasingly, in the bee world, chemical contamination in old combs is a concern.

TIP 239: *Bee defecation is not a hygiene issue*

❀ Much of the interior of the bee's abdomen is empty space. As the foraging bee fills it, the elastic 'crop' bulges into the extra space as it loads with nectar. Alternatively, when confined, the rectum will fill with faecal material and bulge into the extra abdominal cavity. During winter months, the bees will commonly need a day or two per month to take flights to relieve themselves outdoors. Upon release, nucleus or swarms, of bees, having been confined in captivity during transportation for a few days, will have a need to cleanse their system. This is not a hygiene issue: there are no known human diseases spread from faecal matter.

TIP 240: *Feed nucleus or swarms of bees thick syrup regularly*

❀ The spring nectar flow is not always dependable. Warm weather is arriving while cold weather is passing. The nucleus colony will need resources to generate heat for the developing brood nest and to keep adults fed and warm. Many nucleus or swarms, may not have the resources to withstand even a few days of coldness and nectar dearth. During early spring, any colony that is light in stores should be fed, but in particular, swarms and package bees will need feeding. Feed syrup as thick as it can be mixed with hot tap water. Most likely, heating the mixture on the stove is better, but much more work is required. Don't scrimp on the granulated sugar. This is the carbohydrate the hungry bees need.

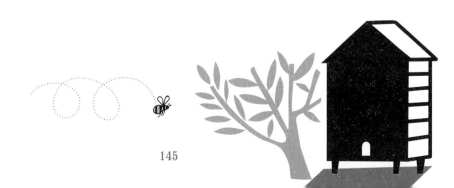

TIP 241: *Keep the queen confined after bees are released*

⚜ Commercial package and queen producers are in the business of just that – producing bees and queens. Very little, if any, commercial pollination work is done and a honey crop can actually be an advantage to the process of colony build-up. The new nectar makes the bees stick to the shaker funnels and may cause the bees. Bees from several colonies can be mixed together to make up the 1.4kg (3lb) needed for an individual nucleus.

At the same time, queens are produced in a different part of the operation at a different location. While one team is collecting the right kind of bees, another crew is 'catching queens'. Late in the afternoon, freshly caught caged queens are placed in newly shaken bee nucleus. Most of the nucleus of bees and the queen are not familiar with each other. If the new queen is released too quickly, many of the bees will still view her as a foreign queen; consequently, the queen should be kept in her shipping cage for 5–6 days, but rarely longer than a week. Older books may recommend the release of the queen in 2–3 days, but that general recommendation has changed. Allow more time than this before releasing her.

TIP 242: *Do expect some queen problems*

🐝 Often caged queens are replaced by the colony within a few weeks of release. This queen superseding event can be the death knell for this now-queenless young colony. The new colony will simply not have time to replace the queen, develop a brood nest and build up necessary adult population and food stores for the upcoming winter.

Frustratingly, it is not uncommon for the replaced queen to look perfectly normal to the beekeeper. If a queen is lost, probably the best solution is to combine the now-queenless new bees with a queen-right colony (see tip 227). Allow the larger colony to settle, and as soon as possible, buy a replacement queen. Approximately two weeks later, make a colony split from the combined colony (see tips 251–256). In reality, you are trying to start the process again, but this time with adult bees, some brood, honey stores and a new queen.

TIP 243: *Keep nucleus colony disturbance to a minimum*

🐝 These new colonies need their privacy and a chance to become familiar with each other. New beekeepers are generally eager to be helpful and to watch the progress of these new colonies, but as much as possible, intrusions should be kept to a minimum. When these new colonies are opened, it cannot be mentioned enough that you should use only a small amount of smoke and keep the examination brief.

TIP 244: *Treat the colony with an approved control for mites*

🐝 This tip may cause some concern for experienced beekeepers who do not routinely use chemicals in their colonies, but a Varroa mite (see tip 179 for more on Varroa mites) control programme is recommended – even for package colonies. It is common to assume that the new colony has no Varroa mite load and that no control programme will be needed. This may or may not be the case. If there are other established colonies in the apiary that are not being treated for mites, this recommendation becomes even more important.

Give these new colonies a chance to become established before requiring them to confront this parasitic mite. Certainly, monitor mite populations as these new colonies develop. At this early stage, they are vulnerable.

PURCHASING COLONIES

TIP 245: *Be responsible if you select large colonies*

Probably the best way for most novices to begin a beekeeping project is to start with a small colony, either a package or a nucleus (see tips 251–252). Both the new beekeeper and the small colony can mature together. But for others, the occasional opportunity presents itself and producing colonies can be purchased. This is not the wrong way to become a beekeeper, but it does put added pressure on the new beekeeper to learn and develop quickly. Primary responsibilities for large hives that must be immediately considered are stinging, swarming, hive management and disease management. As has been discussed elsewhere, neighbours may be a factor taking on such a fast-paced hive-development programme.

TIP 246: *Work with other beekeepers to find colonies for sale*

Within the beekeeping industry, a system has never evolved for systematically providing established, functional beehives for sale. Occasional beekeepers may take on the service, but the size of the colony, the condition of the equipment and the availability dates are all variables that must be considered. Additionally, such hives are in demand. In years past, the classified advertisements in various farm publications would occasionally list colonies for sale, but this is now uncommon. Check on the web, but be cautious and expect to have to pick the colonies up on location. Occasionally fully stocked colonies can be purchased for prices something like £250 upwards per stocked single deep.

TIP 247: *Know what constitutes a 'good' hive*

Hive appearances mean a lot, but not everything. A really powerful, healthy colony can be housed in scruffy equipment. Alternatively, a weak colony can be made to look much better by installing it in clean, freshly painted equipment. Depending on the selling price, each colony may be worth the money. Consider the condition and the precision of the hive – homemade or commercially produced? How much equipment is included? Of course, the bees and the season are important factors to consider. Strong colonies in early spring are generally more valuable than strong colonies about to go into winter.

Established colonies can be tricky to buy. Look for a fair deal and get the opinion of trusted beekeepers. Ask an experienced beekeeper or your local bee inspector to examine the hive with you for any diseases or maladies.

TIP 248: *Get your hive purchase transaction reviewed*

Just as there are no standardised bee colonies for sale, the ability and experience of individual beekeepers also varies. For most of us, keeping bees is an enjoyable endeavour. We strive to be good at the craft, but we know that we will fall short occasionally. If you are brand new to beekeeping, ask a beekeeper with more experience to review the deal you are exploring. Keep in mind that your beekeeping friend's advice and opinions are only his best guesses. Upcoming variables and unknowns such as weather, nectar flow and disease issues can all affect the hive's future – sometimes positively and sometimes negatively.

TIP 249: *Be prepared to transport established colonies*

When purchasing established colonies, be prepared for the responsibility of moving the newly purchased colony to your location. Moving bees is a specialised aspect of beekeeping. Talk to beekeepers who have experience in moving colonies. It is essential to remember that a populous colony can suffocate within 10–15 minutes if screened tightly on hot days. Always allow for abundant ventilation at the entrance and the hive top. Do not be afraid to wet the confined colony with water from a hose. If at all possible, have enough fuel for a round trip. Escaped bees at petrol stations can worry fellow customers. Do not leave the trailer or vehicle parked in the full sun and be fully suited when you release the pent-up bees. Expect newly located bees to be feisty for several days and give them some quiet time to settle in.

TIP 250: *Importantly, be mindful not to buy too many colonies too soon*

🐝 Depending on their personality type, some new beekeepers must learn to control the passion they feel for their new enterprise. The common feeling seems to be that if 1–2 colonies are so very enjoyable, even more colonies would be proportionally more enjoyable. Well…yes and no. As more colonies are acquired, the personality of the bee operation slowly changes from hobby to sideline job. Some beekeepers simply add too much labour too soon to the mix. This responsibility can become tiring. But take it at the right pace and, for many people, keeping bees will become an enjoyable part of their life that will go on for many years.

MAKING NUCLEUS

TIP 251: Search for hive 'nucleus' producers

One common but non-standardised way to initiate a beekeeping project is to start with a colony 'nucleus' or colony divide. Rather than buying an entire hive colony the seller divides a populous colony into several smaller units – generally 3–6 frames – and introduces a new queen. Good nucleus hives are made of brood (all stages), food stores and the new queen. Since this unit is simply a colony nucleus, it is nearly mandatory that nucleus be purchased at a time when a nectar flow is imminent. The small colony will then grow into a full-sized hive. In some instances, you may be expected to provide the hive equipment. Alternatively, corrugated paper board boxes are occasionally used for temporarily transporting the colony.

TIP 252: Know what to look for in a good colony nucleus

Assure yourself that the producer has a good reputation and provides a good product. Did the producer raise the queens or were they purchased from another producer? Purchased queens will probably have been shipped and stored for a while. Before initiating the deal, have a clear understanding of how much and what kind of brood will be included in the unit. Ideally, a small population of both open and sealed brood will be included as well as a frame or two of capped honey. If all goes well, you will need to transfer the nucleus into full-sized equipment. This is imperative for the future wintering success of the developing colony. For nucleus or swarms, the first winter season is a true challenge for existence.

TIP 253: *Plan for larger nucleus to develop faster*

A colony nucleus is a relative thing. It can be large or small. While 3–6 frames make up most splits, a large colony can be divided by hive bodies rather than individual frames. For instance, if a thriving hive is comprised of three deep boxes, those resources could be broken into three large nuclei (10 frames each). The catch will always be the availability of new queens to head these large nuclei. Don't be tempted to make a large nucleus and allow them queen components to produce natural queens (see chapter 5). If divides are made early enough and queens are available to head the new single-story colonies, they will grow and develop rapidly. The brood population is the key. Larger nuclei make full colonies more quickly, but they will require many more bees and brood. If they are being purchased, they will certainly cost more money than smaller nuclei.

TIP 254: *Make nuclei in the autumn*

In warm climates, colony nuclei can be made in the autumn when bees and queens are more plentiful. Everything is performed the same as nucleus made during the spring season. Feeding will be important and possibly honey frames will need to be added during the cold season. For autumn season divides to have a good chance for success, the procedure should only be attempted in areas with very mild winters. The advantage to this procedure is that the colonies that can survive the winter will be ready to build up next spring. Even a nucleus that is weak can build up nicely given the passing of winter weather and abundant food stores. In cold climates, the chances for success are significantly reduced.

TIP 255: *Correct problems or dispose of unsuccessful nucleus*

🐝 It's no surprise that some colony nucleus efforts are not successful. Much of the time, the queen will be the problem, but other problems can be caused by diseases, lack of pollen availability or nectar flows. Traditional colony-management procedures are commonly used to correct the problem. If the queen is present and the issue seems only to be the food supply, then feed the young colony heavy sugar syrup on a regular basis. Only give what the needy colony can process; otherwise, the unused sugar syrup may ferment before the bees can consume it. If the queen is not successfully accepted, it is doubtful the nucleus can be saved unless it was a large one. The problem is the same with failed queens in swarms. Fundamentally, brood production is interrupted to the extent that the colony cannot recover in time to build a foraging bee population and gather enough stores for the next winter. The common procedure would be to combine the failed colony with a successful colony. If you are really new to beekeeping, ask a more experienced beekeeper for advice about this task. Be sure that the successful colony is not damaged by combining the queenless component with it. Don't make a problem where there isn't one.

TIP 256: *Monitor ailments and seek advice if worried*

🐝 You will be the colony's first line of medical defence. Assume that your colony has a Varroa population and monitor its level. Watch for other diseases, especially infectious brood diseases such as American foulbrood. Talk with others at meetings or explore web resources. Select a mite-control programme that fits your management philosophy. It is very easy to become complacent and not treat regularly for Varroa mites. They can build up quickly, so stay ahead of them. (See chapter 7.)

SWARM BEES

TIP 257: *Prepare for the swarm season*

While reasonably common many years ago, today's beekeeper must consider the occasional big swarm as a rare gift. Though such swarms are always enjoyable to hive, new beekeepers should not rely on getting a large swarm as a dependable way to either get into beekeeping or increase colony numbers. These swarms are just too random. Such a large swarm, maybe 1.8–2.2kg (4–5lb) of bees, will normally build up nicely and possibly provide a surplus honey crop the first season. Tell local pest-control operators, county extension offices and possibly the local fire department that you are interesting in getting swarm calls. Tell your friends. Tell fellow office workers. If your local beekeeping group has a swarm list, put your name on it. Good luck. (See tips 167–182 for more on the behaviour of swarm flights.)

TIP 258: *Be prepared for the swarm call*

When the call finally does come, you must move fast. The swarm may not stay very long. The best primary swarms issue during the spring season. Experienced beekeepers have boxes ready to go and already in the boot of the card. A swarm box can be a nucleus box, a deep brood box or even a corrugated board box. A frame or two of drawn comb really improves the attractiveness of the swarm box to house-hunting bees. Swarm lures, with a pleasant citrus odour, are chemical attractants that mimic the primary components of the honeybee's swarm pheromone. These lures may be helpful in enticing the swarm to a bait box. They help, but are not necessary.

TIP 259: *Use old, discarded equipment as 'swarm bait'*

Paper pulp potting containers are available from bee supply companies and serve as swarm boxes. But as many, many beekeepers already know, old, unused equipment that is often carelessly stacked about seems to be highly attractive to swarm scout bees. If you have old equipment that is destined for the burn pile, give them another life as a swarm bait box. If you use full-sized boxes, greatly reduce the entrance and put extra barriers near the entrance. Otherwise, birds and squirrels will take over the box as a nest site for the wrong animal. If possible, put these boxes 1.8–3m (6–10ft) off the ground. The height from the ground doesn't seem critical. Leave the wax and propolis residue within the old box and possibly put a few frames of old combs in the boxes. Be sure the comb is just old and not contaminated with American foulbrood (see tip 297), other diseases or pesticides.

TIP 260: *Know what questions to ask when the call comes*

To prevent wasted trips for disappointing swarms, ask the following questions before leaving. The order for asking these is not important.

- How large is the swarm? A small, mating swarm (also called a secondary swarm) is difficult to capture and slow to build up.
- For sure, are the bees in question actually honeybees? To many people, any wasp, hornet or bee is a honeybee.
- How high off the ground is the swarm? Do you have a ladder? Very high swarms are nearly impossible to lure to the ground.
- How long has the swarm been at the site? Some swarms that have gone queenless for a while can be bad tempered.
- Does the caller have the authority to give the swarm to you? If relationships are not the greatest, the occasional caller may be eager to get rid of the neighbouring beekeeper's bees.

TIP 261: *It is OK to say 'no' to a swarm offer*

For the passionate beekeeper, nothing is more distasteful than giving up on hiving a swarm. Maybe the swarm was too high or some of the bees seemed to already have begun moving into a cavity in the wall of a building. Swarms that have already begun occupying a wall cavity cannot be hived using normal swarm-hiving techniques. Some of the swarm calls that beekeepers get involve bees that are simply not accessible. Though the beekeeper will feel a sense of guilt about leaving the swarm unhived, some swarms just are not worth it and you just have to say no. Scouts from the unhived swarm may find another nest cavity or the swarm members may try to build exposed combs. An exposed nest open to the elements is rarely successful.

TIP 262: *Be careful with any bee-retrieval missions*

Having just discussed 'learning to say no', to the occasional swarm caller, many beekeepers do attempt superhuman retrieval techniques! Don't climb high ladders. Don't try to hose them down with water and no, you can't use a firearm to shoot either the limb or the swarm.

Keep in mind that the actual monetary value of the swarm is probably around £30-50. Be clever, but don't be foolish. The biological problem is that the high swarms – even if dislodged or shaken by whatever clever contraption the beekeeper has devised – will fluidly and gracefully take flight and more than likely simple reperch on the same high limb. If the queen flew high and landed on a branch, a powerful odour cue remains there. Rather than move down to the ground, most swarms will simply move back to the high branch bearing the queen's chemical mark.

TIP 263: *Always approach the swarm with a bit of caution*

🐝 It is true that the occasional beekeeper has been given the famous '15 minutes of fame' due to a swarm landing in a highly visible and public place. However, any beekeeper at any time should approach the swarm with respect and caution. Have your veil close at hand. The typical new swarm is gentle and agreeable – but not always. When the swarm departed, individual members took food with them. After all, they didn't expect to be returning to the original site. But suppose a cold rain fell before the swarm could occupy a new nest site. A swarm that has become trapped in its temporary location may have been forced to consume its temporary food supply. Such a swarm becomes cold, hungry and cross. If a beekeeper just walks up to this swarm without proper protection, significant defensive retribution should be expected. Until you know the status of the swarm, show respect and caution.

TIP 264: *Learn how to handle a 'dry' swarm*

🐝 When you approach any swarm, the bees should essentially ignore you. Alternatively, if you approach the swarm and the individual bees are runny and skittish and if a bee or two or three seems to want to sting you, give this swarm some room and respect. Gently dribble thick sugar syrup over this testy swarm for 30–45 minutes. Feeding the hungry swarm can change its attitude, making it more manageable. Remember, the beekeeper should always be in charge. Don't make a bad impression with the public and don't lose control of the swarm.

TIP 265: *Before proceeding, search for the queen*

To the uninitiated, finding the queen in the moving blur of several thousand worker bees would seem completely impossible. Admittedly, some skill and a good deal of luck are needed, but many times, the experienced beekeeper will be able to spot the queen. In order to fly, she will have lost weight and will be on the run on the surface of the swarm. Try to grab her by either her thorax or her wings, but do not hold her by her abdomen or her head. Don't use gloves. It is not as difficult to pick up a queen as you might expect. Confining the captured queen in a cage will nearly guarantee the swarm will stay in your hive box (see chapter 4 for more on queen bee and colony behaviour).

TIP 266: *Direct a swarm into your bee box*

After having a look for the queen and finding her – or not – the next step is to get as much of the swarm in your hive box. Every swarm situation will be different. Set the box near the swarm. If the branch is small, snip it off and move the swarm and branch to the box. The bees will begin to move into your equipment. If the branch can't readily be cut, do whatever it takes to get the box near the limb and give it a sharp jerk. Bees will fly everywhere, but if you have an attractive box with an old comb or two in it, no doubt some of the flying bees will find it. Watch for bees scenting at the entrance of your equipment. That's a very good sign that the bees are interested.

TIP 267: *Be patient when it comes to getting the swarm*

Clearly, some swarms are very difficult to hive and even if you do get the bees, they will not always stay in your box. Be patient and methodical. Always remember that sometimes things just don't work out and you may get better luck on your next attempt.

TIP 268: *Old folklore isn't your strongest bet*

An urban legend that is alive and well in bee lore is that bees can be 'tanged' down, also referred to as a 'tangle of bees'. The procedure is for the beekeeper to bang two pieces of metal together to entice the swarm to land. Apparently, this process seems to have derived from beekeepers who would cross property boundaries while chasing a swarm that issued from their skeps. The commotion was to alert others that a swarm was in transit and that the chase was on. At this point in our scientific understanding, there seems to be no behaviour reason for bees to settle just because of the sound of metal being clanged; however, there is no harm done in trying the old technique.

TIP 269: *Understand that it might be a small swarm*

Be prepared for the disappointment of a small swarm. To many concerned homeowners, all swarms appear large. These small swarms may be called after swarms, mating swarms or secondary swarms. The swarm will be made up of about 0.22–0.45kg (½–1lb) of bees and are usually headed by a virgin queen, which makes the small swarm unstable and skittish. Such a swarm is difficult to hive and may not build up enough to survive the winter. However, if the queen does begin to lay eggs, certainly that queen could be used to replace other failing queens. A novel use for these small swarms is teaching children the fundamentals of beekeeping.

BEE REMOVAL

TIP 270: *Expect each bee removal job to be different*

🐝 Bees in the wrong place can be a problem. When searching for a home in nature, scout bees look for a nest cavity that is about one cubic foot in size, and that is dark and dry, with a defendable entrance and nothing else living there. As forests have disappeared and old trees with cavities that could serve as a nest cavity have become rarer, scout bees long ago began exploring cavities in buildings and houses. Many times, there is not a problem, but if they are near a doorway or if they find their way into the house, they can become a pest. Before attempting to remove bees from other people's property, know what you are doing.

TIP 271: *Ask for compensation for difficult removals*

🐝 Information is readily available giving instructions on ways to get these bees out of house cavities. Just as with swarms, some colonies in houses are easier to remove than others. Two broad plans exist for dealing with these bees. One way is to trap the bees out with cone traps. This is a slow procedure that will take much of the summer. A simple screen cone is used to let bees out of the nest, but will prevent them from getting back inside. The second common way is to cut an opening in the house and remove the colony. This is a significant procedure to undertake. Thousands of bees will be flying around and honey will be dripping from broken combs. Surprisingly, the bees will quickly become disoriented and confused. There will be some stinging bees, but not great numbers. Accumulate experience with another beekeeper who has been through this before. Beekeepers who are specialised in this carpentry procedure charge meaningful money for this task.

TIP 272: *Educate the building owner about nest removal*

Too often the property owner thinks that the value of the honey and bees will be enough to cover removal costs. That is rarely the case. Indeed, the occasional owner may even suggest that some part of the honey crop should come their way. The reality is that colonies in house walls tend to be much smaller and have smaller honey crops than typical managed beehives. If the project is not undertaken during the correct season, the colony will not survive the upcoming winter. There are no magical chemicals that will force the bees out of the nest and smoke is of little use other than for simply subduing the colony. The only way to thoroughly solve the problem is to cut open the building and remove bees and combs. If the bees are killed with insecticides in the nest cavity, a few days later the stench of decay will become evident and honeycombs will potentially break and release honey in the cavity walls. Adding to the confusion, mice, ants and other vermin will be attracted to the area.

TIP 273: *Leave the high jobs to better equipped keepers*

Bees can sometimes find nest cavities very high off the ground. Nests located in such inaccessible places may not be a problem until repair work must be done. Even workers who are accustomed to lofty places decline jobs that involve bees. Specialised beekeepers with specialised equipment such as vans or scaffolding are needed. Leave the high work to those with the expertise to do the work.

TIP 274: *Anticipate a difficult removal from brick buildings*

Generally, bricked structures, either solid brick walls or brick façades, present more challenging issues than a wood-framed structure. Spaced several feet apart, holes along the bottom edge of the wall provide common places for bees to enter. These intentional holes provide an opening to allow water to drain from the bottom of the wall and also provide for ventilation behind the wall to help keep the structure dry. Bees and other insects can find their way into the house through these openings.

Additionally, where trim and fascia join the brick wall, only a small opening is required for scout bees to find an entrance. As with other problem honeybee nests, the bees cannot be smoked out or forced out with repellents. Be prepared to have to remove the bees from inside the brick house rather than the outside. It is worth making sure that your beekeeping insurance covers this type of work.

TIP 275: *Look beyond the entrance for an established nest*

A major challenge that can occur with a nest removal project is that the bee nest can be far away from the entrance where you see them going into the wall. This can result in two damaged wall sections: one to open the wall for a look and the second spot nearer the nest location. There is no easy way to confidently check this situation, but one clue may be the warmth the nest gives off. Sometimes the warmth of the nest area can be felt through an inside wall with the palm of your hand.

A doctor's stethoscope is useful for listening for the nest's buzz through the wall. High-quality devices are pricey, economy models are available on the web. Though interesting and occasionally useful, these devices would only be useful to serious bee removers.

TIP 276: *Use specialised equipment to remove difficult nests*

🐝 A flexible video camera with a USB connection has been useful to specialised beekeepers who are removing bees from buildings. Depending on how near the entrance the nest is located, a borescope can be fed into the entrance while the image is projected onto a computer screen. Desirable models of this equipment have LED lights that illuminate dark areas. These devices are used in the plumbing industry, while specialised models have medical uses. Surprisingly, many models are inexpensive. Currently, there is not a specific model that is useful to beekeepers, but these gadgets would be useful to the bee nest remover.

TIP 277: *Use a bee vacuum for serious bee removals*

🐝 A bee vacuum is a vacuum device that collects honeybees in a trap in the suction line rather than in the vacuum canister. Using a shop vacuum without the trap will kill bees in the tornadic air flow inside the vacuum tank. Bee vacuum units are available from commercial bee supply companies, or several different plans for shop-built models are readily available by simply searching 'bee vacuums' on the web. In addition to shop vacuums, beekeepers have successfully modified leaf blowers and gasoline-powered bee blowers to provide portable air supply sources.

Regardless of the model chosen, a vacuum air speed will need to be selected that will just-ever-so-slightly pull clinging bees from combs. If heavy vacuum force is used – a force so great that bees are immediately sucked from combs – too many bees will be killed in the tubing as they tumble towards the trap. This vacuum device is not a common piece of equipment but is a must for beekeepers who retrieve difficult swarms and remove bees from the honey house.

TIP 278: *Stay alert for bee trees that are already down*

Established beekeepers can expect occasional calls from homeowners who have a honeybee nest in a tree cavity. Interestingly, these natural nest cavities can be quite high from the ground, but a fear of stings and a lack of understanding will force the owner to want them removed. Much of the time, the owner will not allow the tree to be damaged. The beekeeper must decide if this job is worth the effort. As with removing bees from a dwelling, removing bees from some tree nests is simply not worth the effort. If the tree is down and can be opened, the task is much easier and can normally be accomplished using traditional techniques.

TIP 279: *Use an exit trap on living trees*

If the tree is alive and standing, and cannot be damaged, the only technique for removing the bees is with an exit trap. Such traps are simple cone-shaped gadgets that are stapled over the entrance. The tip of the cone is open a bit – much like a funnel – allowing the bees to escape, but not to easily return, similar to clearing a super. A functional beehive should be positioned nearby. As bees accumulate and are unable to get back to their nest, they often adopt the nearby hive as the only alternative. This procedure will take weeks to accomplish.

NEST TRANSFERS

TIP 280: *Learn to transfer combs from the natural nest*

Once the nest cavity is opened, the process of transferring combs to traditional frames is simple. Using a knife, cut brood combs into rectangular shapes and fit them into an empty frame. A lot of space and rough fits will be the result. Using twine or large rubber bands, tie the comb pieces into the frame. The suspended combs will be loose and wobbly. This process will not be neat and it will not be pretty, but when given to the bees, they will attach and firm up the loose comb pieces overnight. After the brood has emerged from the combs, the rough comb pieces should be removed from the colony. These frames are only to prevent waste of colony resources and to provide an attractant to the bees during the transfer process.

TIP 281: *Ideally, transfer colonies during spring months*

The transferring process should only be undertaken during spring or summer. This is a radical process for the transferred nest and even if some of the brood and honey stores are saved, the colony still must rebuild combs and resources before the upcoming winter. If colonies are transferred later in the summer or autumn, bees and nest resources should be combined with another colony. Otherwise, the transferred bees are certainly doomed.

TIP 282: *Transfer an exposed nest as early as possible*

In nearly all instances, an exposed nest will die even during a mild winter. Such nests can readily be transferred to traditional hive equipment. Colonies transferred during the spring have a decent chance of survival, while colonies transferred during late summer will need lots of help.

AILMENTS of HONEYBEES

Keeping colonies vigorous and perpetually healthy is an elusive goal, not a plan. Though bees have excellent natural health systems, they can become ill and frequently need our help. The beekeeper is the colony's first line of assistance, though be mindful that we can only help some of the time.

EVALUATING COLONY DISEASE

TIP 283: *Occasionally, bad things happen to good colonies*

Today's beekeepers, when talking to older colleagues, are frequently left with the impression that all was fine in beekeeping before Varroa. That impression is both right and wrong. Things were different and even simpler, but even before the presence of Varroa, beekeepers' colonies were challenged. Mysterious events seem to befall colonies both then and now. It is not that rare for something to go wrong with a good colony. It's not always diseases. Storms and flooding can destroy colonies. Rarely, but sometimes, vandalism is an issue. Having things go wrong occasionally should be expected. The best colony will only be the best colony for a few years. Colony quality is always waxing and waning. It is a natural scheme.

TIP 284: ... And most colonies are normally healthy

While bad things do occasionally happen to good colonies, it should be noted that most colonies are generally healthy – so long as mite populations are suppressed. But by being generally healthy, don't think that means generally exceptional. Most colonies can usually be ranked average to good. Exceptional colonies are much rarer. One of the recent shifts in bee colony management schemes is that today's beekeeper must be more involved in keeping the colony generally healthy, or within a few years, average colonies will slip into the failing category. Expect most colonies to be average but enjoy the occasional exceptional colony. More than likely, this outstanding colony will fade within a few seasons, but other colonies may very well rise to the top.

TIP 285: ... But do cope with ailing colony frustration

New beekeepers should expect to experience occasional frustrations with their colonies. While beekeeping is presently deeply satisfying to many people, when bee things go wrong, the disappointment can be palpable. At one time or another, nearly all beekeepers have felt these emotions. Investment money seems lost. Honey flows do not materialise. Swarms escape. Queens fail. The list can be long and depressing. If beekeeping was easy and success was guaranteed, nearly all of us would be beekeepers. Dealing with disappointment and frustration is a normal aspect of the craft. It's the same way with gardening or farming. When the occasional bee loss hits, accept the disappointment and loss, and regroup. Know that many others have already experienced this and that others should plan for this type of event to happen. Beyond the setback, your beekeeping time will brighten again.

TIP 286: Know the Government regulations on diseased colonies

Within the EU, all regulations for beekeepers are the same. All information on regulations towards beekeeping can be found on the Bee Base programme on the DEFRA website (see page 288 for information). Bee Base enlists information on all legislation within the EU for diseases and importation stating that American foulbrood (AFB) and European foulbrood (EFB) the Small hive beetle (*Aethina tumida*) and Tropilaelaps mites as notifiable pests and diseases throughout the EU. Imports of honeybees (*Apis mellifera*) will only be allowed into the Community from third countries listed. Provided that the three notifiable diseases of bees in the EU – American foul brood disease (AFB); *Aethina tumida* (small hive beetle); and *Tropilaelaps spp.* (Tropilaelaps mites) – are also confirmed as notifiable diseases throughout the exporting country. The bees must come from an area that is not affected by these notifiable diseases. The bees must have been inspected and certified as being free of diseases, including notifiable diseases and infestations affecting bees.

TIP 287: Learn to assess a colony's general health

The season of the year, the level of entrance activity, the number of pollen carriers and the number of dead bees and crawling bees at the entrance, combined with a quick estimation of the colony's gross weight, can give the experienced beekeeper a quick idea of the colony's present condition. But the evaluation is more than that single observation. What are other colonies in the apiary doing? Is there a general sense of vigour and activity or is the apiary a bit quiet? Are flying bees going about their business or are a few bees dive-bombing those who visit the apiary? Sometimes these general hive evaluations are wrong, but much of the time, all the beekeeper needs to know can be acquired just by giving the apiary scene a good look.

TIP 288: *The source of a problem is not easy to spot*

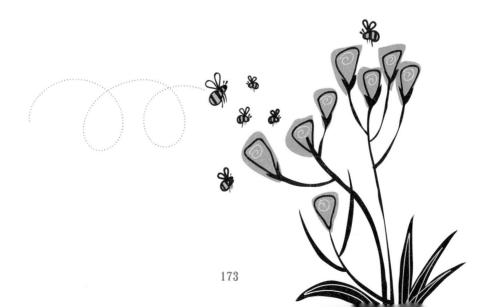 Foraging bees are wide-ranging animals. Consequently, knowing exactly where bees have just been is presently impossible to know. Occasionally, a colony comes down with an ailment and the beekeeper is curious to know where it came from. If the beekeeper just bought used equipment or bees from another beekeeper, no doubt that transaction will be suspect. But many times, there will have been no unusual involvement of other bees or beekeepers. The source of a particular malady may never be known. Indeed, the infection could have come from your own bees.

Bees' immune systems are constantly fighting infections. If colony stresses, due to weather, water shortage, pesticides or queenlessness, are severe enough, it could be that an ailment that had been previously contained by the bees suddenly becomes apparent. On occasion, a beekeeper's colonies become infected with something like American foulbrood (see tip 297) and a neighbouring beekeeper is wrongly blamed or at least suspected. Sometimes, the source of the infection is simply not known.

TIP 289: *Adopt a plan for helping future ailing colonies*

🐝 In the past, only a couple of ailments had potential chemical remedies, but modern beekeepers have a host of possibilities for controlling, or mitigating, common diseases. Each beekeeper must now decide what their colony treatment philosophy will be. Many beekeepers will opt not to use any chemical concoctions at all, while others will readily apply medications that promise to improve colony health. While Apivar and Apistan are viable options for some beekeepers, others will desire nonchemical or soft chemical treatment programmes or even no chemicals at all. Products like Apivar and Apistan are chemical products and require care when using as a treatment method. Still other beekeepers will use chemicals from either category depending on the severity of the infection. Clearly this is a personal philosophy for each beekeeper. You will need to decide yours.

TIP 290: *... And develop a long-term plan*

🐝 Normally, assisting ailing colonies is an unclear process that requires time to show improvement. The various treatments used will be predicated by the medication viewpoint of the individual beekeeper. Whatever treatment scheme is selected must be consistently employed. Some diseases, such as American foulbrood, a persistent bacterial infection (see tip 297), will require a distinct programme for eradication and adherence to Government regulations, which is to inform the authorities immediately and do not move the affected colonies until government inspectors have inspected the hive. Select a philosophy and then select a plan.

TIP 291: Realise that a plan will reassure nearby beekeepers

✿ In reality, little time is actually spent within the beehive, but much more time is spent preparing for the open hive event or talking to other beekeepers about what transpired within the open hive. By far, most beekeeping time is spent talking and preparing. It is reassuring to neighbouring beekeepers that honeybee diseases and pests are respected and that a disease-management programme is in place. To a greater or lesser degree, all beekeepers are a bit cautious when it comes to bee diseases and their mysterious spread. Everyone wants healthy bees and no one wants bees exposed to diseases within other bee operations. Be a good beehive manager and keep colonies as disease-free as possible. Respect in beekeeper circles and a good reputation will grow over time.

TIP 292: Remember, multiple disease and pest conditions can exist together

✿ Sometimes the colony diagnosis can be made more complex if there are multiple issues within one colony. Most often, one of the diseases or issues will establish itself as the primary cause of the colony's decline. For instance, an exposure to pesticides may stress a colony and allow Chalkbrood to become established. Little can be done for either issue, but helping the colony recover lost population will help the colony re-establish brood nest conditions, and the Chalkbrood condition, which is a fungal disease (see tip 320), may subside. However, if the queen's progeny is genetically susceptible to Chalkbrood, recovery may not occur. In this case, requeening would be the most effective solution. Time will be required to determine whether the treatment procedure is effective. Don't despair. Multiple issues are not common, but they are possible.

TIP 293: *Stay current on new treatment recommendations*

🐝 Staying abreast of control recommendations is easier said than done. Presently there are several schools of thought and many different control materials and procedures. None of the current materials offer perfect control of any bee disease. Varroa control programmes are geared to control approximately 50% of the mite population within the colony. Destroying all the mites is not easily done without damage to the honeybee hosts. Practically all of the available materials work to some extent, but none are head and shoulders above the others. In the case of Varroa control programmes, many beekeepers do not use any chemical control for Varrosis, but for most beekeepers that is a goal and not a plan.

In light of the complexity of control programmes, most beekeepers select a suitable plan that fits with their disease control philosophy. They use this plan until they sense that Varroa is adapting to it or until they learn of some new product that could work better. Current information is acquired at meetings, from technical publications obtained from the British Beekeeping Association (BBKA) and the Department for Environment, Food and Rural Affairs (DEFRA). For now, determine what experienced beekeepers in your group are doing and explore the literature for suitable control products. A beekeeper armed with current information will make better decisions.

TIP 294: *Judge your queen fairly*

🐝 The queen will always be in the hive's hot seat. She commonly gets all the credit for anything good and all the blame for anything bad within the colony. Even though she is sometimes unfairly maligned, sometimes she is truly the cause of a poorly producing colony or a colony with susceptibility to various diseases (see tip 301).

BEE BROOD ILLNESS

TIP 295: *Monitor brood and bees for healthy appearance*

During every colony inspection, monitor brood and bees for a healthy appearance. At first, this is a task that new beekeepers must learn to do, but experienced beekeepers would have difficulty in not doing it. Scanning the brood and bees for problems will become totally second nature. For example, as you open a colony, you will automatically check the area just in front of the landing board. Once inside, you will instantly make decisions concerning the queen by looking at her brood production pattern. As you gain experience, it will become apparent what symptoms are important.

TIP 296: *Learn the basic indicators of brood diseases*

Superficially monitoring brood health may seem difficult, but in short order, the general health of the brood will become obvious to the knowledgeable beekeeper. The brood nest should be clearly defined on 4–8 frames and not helter-skelter throughout the colony.

Some characteristics of healthy brood are: compact egg pattern, brilliant white larvae in a concentric circular pattern within the brood area, uniform tawny-coloured and slightly raised brood cappings, and finally the presence of recently emerged young workers throughout the brood nest. Cappings that are dark and oily are not good indicators. Spotty brood patterns with many random unfilled cells indicate a problem. Dry larvae that are yellowish with brownish smears are not healthy larvae. Hard, chalky-white brood with black smears is an indication of a fungal disease. Any larval or pupal stage on the landing board on the ground in front of the colony signifies a pathogenic issue within the colony.

TIP 297: *Diligently watch for punctured, greasy cappings*

American foulbrood (AFB) and European Foulbrood (EFB) are bacterial diseases of honeybee brood that occur in the UK and are found worldwide. Both diseases are notifiable under the Bee Diseases and Pest Control Order 2006. Foulbrood is caused by a Gram-positive spore-forming bacterium *(Paenibacillus larvae)* that primarily infects three-day-old worker larvae, causing death to occur in later stages. Dark, greasy cappings that are erratic and partially opened are the hallmark of an advanced case of foulbrood. Body fluids from the dead pupae seep into the cappings, giving them a dark, greasy look accompanied by a putrid odour.

As hygienic house bees perceive the dead brood, they begin the process of opening the cells for bee corpse removal; hence the erratically, torn cappings. In undertaking this task, these cell-cleaning bees themselves become carriers of the contagious spores and unintentionally feed AFB-laced brood food to young bees. Though adult bees (and humans) are not affected by the disease, it spreads to other young larvae within the colony. An important point is that as bees cap healthy pupae, there will briefly be a neat, small opening in the centre of the brood cell. This is normal and is not pathogenic. Irregularly opened, dark oily cells are not normal. Also, while the putrefying odour is noticeably repugnant to most beekeepers, some people simply cannot sense the odour. Do not use smell alone.

Though well adapted to honeybees, in most cases AFB is not particularly widespread. But once inside an apiary, if not deliberately dealt with, this disease will spread to other nearby colonies. Though antibiotics suppress the vegetative stages, spores are unaffected. The disease will reappear when contaminated equipment is reused. The accepted final solution is to destroy the bees and honey stores and burn the equipment. Though bees will probably never have the disease, diligently watch for punctured, greasy cappings. They could mean American foulbrood.

TIP 298: *Note any putrid odours*

🐝 Piles of dead bees within and in front of a colony will obviously emit a decaying odour. That smell should immediately tell the beekeeper that a serious problem is at hand. Pesticide kills and diseases caused by AFB or mites are common causes of decomposition odours. Obviously, determine the cause and remove the rotting mass.

TIP 299: *Use boards to monitor Varroa levels*

🐝 Today's beekeeper can confidently assume that Varroa (see tip 244) is well established in his or her colonies. Screen bottom boards or sticky boards are used today not to determine if Varroa is present – it is – but to determine Varroa population levels within the colony.

A general trigger level would be something like 50-plus mites on a sticky sheet per 24 hours with no other treatment in place. But probably a better, more dependable, procedure would be to systematically treat in the spring and autumn with a low-impact material or procedure. What that material or procedure is would have to be your call. Any hope of finding a chemical fix that will eradicate Varroa is not on the horizon. For the foreseeable future, beekeepers will have to accept a continual level of Varroa infestation within all their colonies and probably have to use a product or procedure to keep mite populations below economic thresholds.

TIP 300: *Do your utmost to reduce mite levels*

🐝 Recently in the beekeeping world, there have been colonies where nothing is being done to control Varroa levels and bees seem OK. Certainly, there is strong hope that these colonies remain (apparently) unaffected by mite predation, but at this time, no one knows.

More commonly, other groups of beekeepers are constantly losing their bees to a host of ailments, many of which are caused by Varroa predation. These beekeepers routinely have to treat with various materials to keep populations suppressed. For most beekeepers, doing nothing is a true gamble. Without good numbers on who is treating and not treating, possibly beekeepers with small numbers are the chance takers. It would seem unlikely that a beekeeper with a significant monetary and energy investment in their bee operation would be eager to take the same chance that a beekeeper with fewer colonies is sometimes willing to take.

At present, most beekeepers should expect to have to use a treatment procedure to suppress mite populations. In fact, many beekeepers select two procedures and then rotate them on a seasonal basis.

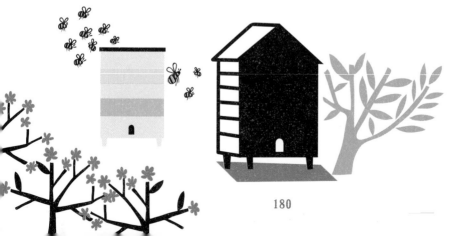

TIP 301: *Know that queen issues are not always pathological*

Blaming the queen for a colony's disease is a tricky area. The queen's offspring may have been made genetically susceptible to a disease or the inoculation could have been so intense that the colony succumbed to a particular disease. Alternatively, a poorly mated queen may simply be unable to produce enough worker brood to sustain the colony's needs. Ironically, removing the queen usually provides a brood rearing break and allows the house bees to catch up with the infection. Even if the queen is not at fault, her replacement may help to resolve some disease issues.

TIP 302: *Be cautious when making nucleus or mixing equipment and combs*

Beekeepers are the main route by which most honeybee diseases are spread. Be able to recognise American foulbrood and realise that increasingly, apicultural scientists are concerned that old combs may harbour disease. Many times, a beekeeping associations may have a small, testing device that members can use. This is a valuable service, but be sure the equipment has been thoroughly cleaned to avoid any transmission of American foulbrood, a bacterial disease of honeybees. Dysentery and other digestive diseases can be transmitted via old combs too. Possibly, writing the date on frames and replacing beeswax combs about every 3–6 years will be a good start. Some beekeepers are replacing a few combs periodically rather than replacing all combs at once.

ANIMAL PESTS

TIP 303: *Be cautious of vertebrate pests and animals*

🐝 For the most part, vertebrate pests are only annoying and are not of great consequence. Once an apiary is found, over a few nights animals can destroy and scatter all the hives in an apiary. For most beekeepers, with the hives left on open farmland, cows are known to be curious and will rub up against the hives. This will cause damage and may even knock the hives over or, worse, upset the bees which may then sting the animals with dire consequences. Many colonies are left on the moors where sheep graze and in a similar way, sheep have been known to cause damage to hives. For example, placing hives near a stone wall can cause problems as sheep tend to jump over walls, therefore knocking hives over. Be aware of openings in the walls and don't put hives too close to walls.

Otherwise, smaller animals such as mice, birds and slugs are annoying to both the bees and the beekeeper, but probably minimal disturbance is done to the colony. In the UK, squirrels rarely engage a populated colony, but they will cut holes in stacked equipment and construct nests inside empty equipment. If you live in the vicinity of woodpeckers they can become a nuisance in damaging the hives by drilling into the sides mostly the brood chamber area. Woodpeckers will occasionally select colonies within an apiary and literally cut them to pieces. This does not happen often, but when it does it is costly to the beekeepers and disruptive to the bees.

TIP 304: *Watch for signs of such pests*

🐝 Telltale signs of woodpeckers are the scratching or small or large holes on the side of the beehive. The best preventative course of action is to wrap chicken wire or strong metal around the sides. Domestic animals, such as sheep, freely roam on large farmland and tend to rub up against objects. Free standing hives are the perfect "scratch fork" for them. Be aware of loose sheep wool and droppings around the hives. Slugs are also a nuisance, so look out for slime trails along the sides of the hives or on the landing boards (see tip 306). Mice are another pesky pest who look for shelter, warm and especially food (see tip 305). Again, watch for signs of droppings or gnawing at the woodwork. One of the most serious pests to watch for is the Varroa mite. Keep a close check on the mite drop and follow your Integrated Pest Management (IPM) method.

TIP 305: *Help the colony resist mouse invasions*

🐝 Modern-day hive entrances use various means to reduce the height of the entrance to no more than 1 cm (⅜ in). Generally, a mouse cannot get through an entrance that shallow. The more common entrance height (2cm / ¾ in) is readily accessible to mice and should only be offered to the bees during warmer months. Mice are only pests when the bees are unable to leave the cluster: during late autumn, winter and early spring. As soon as the weather warms, the bees become alert and will readily defend the hive. Oddly, if the mouse nest is located within the hive but away from the bees' brood nest, mice and bees can live together harmoniously. As the season progresses, the mouse will abandon the beehive and live outdoors.

Mice cut the combs, damage frames and keep the cluster needlessly agitated during the winter months. Mice also relieve themselves within their nest and the beehive. The odour and the filth require both the beekeeper and the bees to perform cleaning that would not ordinarily be required.

TIP 306: *Decide what to do about a pesky pest*

🐝 Slugs are a beekeepers nightmare in the winter but don't seem to do much damage to the bees; however, they create an awful mess with their defacation. They seem to snuggle in the corners of the top of the hive just under the crown board and the corners of the floor board at the bottom of the hive. They have also been noticed to eat honey from the stores. Wherever they gather, the slugs seem to attract a noticeable dampness which itself penetrates into the woodwork. Provide beer traps by putting a saucer of beer down under the hive to attract the slugs. Then gather the slugs daily, which can be labour intensive. Orange or grapefruit peel is also effective.

There are a few natural remedies on the markets for keeping slugs at bay; sheep wool and pumice are just a few. There are quite a few chemicals on the market for slug repellents but one has to be careful that bees, birds and other wild life do not get harmed. Most gardeners' answer is to use a strand of copper wire wrapped around the legs of the hive or hive stand. Slugs do not like crossing copper wire due to its voltaic nature. During the autumn and the winter months where the slugs seem to head for the hives more, make the mouse guard, or a portcullis, out of copper nails spaced 9mm apart.

TIP 307: *Check regulations with councils, land owners etc*

🐝 In the UK there are few larger wild mammals and those that we do have do not cause too many problems. However there are still regulations on keeping bees, and most are common sense. The main one is, 'if bees cause a nuisance to the neighbours they must be removed away from that vicinity'. Bees are classed as live stock and share the same regulations as domesticated farm animals held. Therefore, it would be prudent for the beekeeper to make sure they have insurance for keeping bee, especially if keeping bees on other peoples' property or taking them to events around the country.

TIP 308: *Ants will probably exist in and around the colony*

🐝 Ants and bees, both being hymenopterous insects, are closely related and have a similar physiology; therefore, any insecticides that kill or repel ants will likely kill or repel honeybees. Carpenter ants can tunnel within the hive walls, effectively destroying the hive equipment and ultimately displacing bees from the equipment. But in general, ants stay on the inner cover or in isolated parts of the nest and appear to lead a separate life from the bees. Ant predation should be severe before undertaking extensive measures to prevent ants hive entry.

TIP 309: *The honeycomb moth can cause damage*

🐝 The greater wax moth or honeycomb moth (*Galleria mellonella*) and its close relative, the lesser wax moth (*Achroia grisella*), is found in most parts of the world. They can both become a real pest and natural scavengers of honeybee comb. The wax moth will very easily introduce itself to weakened colonies, entering an empty hive or stored supers to set up home. Damage occurs when the larvae burrow into the comb, feeding on wax, larvae skins, pollen and honey. As they chew, they tunnel through the cell walls, lining them with silk. The silk thread can tether the emerging bee and they starve to death, the phenomenon known as Galleriasis. Wax moth can damage the woodwork of hives and frames. Stored equipment can be protected by leaving them out to the light or fumigating. Certain claims have been made about biological larvicide to control wax moth. Another successful treatment is fumigation using sulphur dioxide strips placed on top of the hives or supers.

KNOW the STINGERS and BITERS

TIP 310: *Be aware of Africanised honeybees*

One, or the main, reason for the ban on importation of queens from the US to the EU is due to the queens carrying the Africanised honeybee trait. Africanised honeybees (AHB) are much more defensive than European bee stocks. Per season, the AHB tends to produce many small swarms rather than the one or two larger swarms typically produced by managed colonies. Due to increased swarming production, the AHB tends to colonise an area and tie up the available natural nesting cavities. Since this bee stock has a tropical heritage, it will more readily nest in open areas or in cavities not normally selected by European stock.

Other than being about 20% smaller and possibly a bit more jittery, the AHB looks nearly identical to the common honeybee of European heritage. On a bee-to-blossom relationship, the AHB is a pollinating insect just as European bees, but this race of bees stings too often and abscond too readily to be of use in commercial apiaries. If you are new to beekeeping and are in an area in which the AHB is established, be sure to work with knowledgeable beekeepers and do not hive swarms of unknown origin.

TIP 311: *Aggressive bees are not always Africanised stock*

Even managed bees can have bad days. These defensive events are nothing that an experienced beekeeper cannot manage, but individuals without protective gear or tethered animals can be in harm's way. If you find that this behaviour has happened more than once, this colony is a candidate for a requeening process, especially if the hive is in a suburban setting.

TIP 312: Asian Hornets are a cause for concern

Occasionally on nature television, Japanese giant hornets (*Vespa mandarinia japonica*) have been shown attacking European honeybee colonies in Japan. In three hours, a group of 30 giant hornets can kill a honeybee colony made of 30,000 individuals. European hornet (*Vespa crabro*) has been present in the Southern counties of the UK for some time. European hornet generally eats live insects, fruit pulp or nectar. Though large, neither of these insects is particularly aggressive and these large insects are not the dreaded Japanese giant hornet that feeds on honeybees. However, a cause for concern is the sightings of the of the Asian hornet (*Vespa velutina*), an invasive non-native hornet from Asia. Its diet is honeybees and other pollinators. Its first arrival to the EU was in 2005 in a containment of imported pottery from China to France. It now seems to have crossed the English Channel to mainland UK.

TIP 313: Watch for pests in remote bee apiary locations

Occasionally, insects and animals turn up in apiary that are not particularly harmful to bees. Beside the odd nest of earwigs or woodlice that don't really bother any bees or beekeepers, mice can give the beekeeper a fright on opening a hive. Stand alone hives or hive on the moors can be at risk during the end of July or August from marauding wasps on the look out for a free meal. Wasps are frequently known to wipe out a colony of bees through their relentless and persistent robbing, causing bees to go on a defensive attack leaving many dead in its wake. The colony will eventually suffer the consequences. The only course of action is to close down the entrance to one bee space. Place wasp traps around or under the hive, and finally, the last resource is to move the hive to a new location.

'INVISIBLE' PROBLEMS

TIP 314: *Spot the signs of burrowing wax moth larvae*

After keeping bees for a few years, most beekeepers will be able to readily recognise wax moths and the damage they cause. These insects are natural degraders of weakened or dead beehives. They become significant in managed bees when both colonies and reserve equipment are readily available for wax moth infestations to become established. While features of comb destruction are obvious, a lesser-known type of destruction is called Galleriasis.

As moth larvae tunnel through the combs, they line the tunnels with silken thread. If necessary, they will burrow through developing pupae. If the afflicted pupae are not killed by the process, in many cases they are tethered to their developmental cell by the moth's silk-lined tunnel and are unable to emerge normally. Since they cannot escape and solicit food, they soon die. A characteristic of the condition will be a patch of uncapped, fully mature worker bees that do not leave their cells. Using a pocket knife blade to remove a restrained pupa will reveal the silk threads and the damage caused by the moth. Since the moth is protected deep within the comb tunnel it has built, worker bees have a great deal of trouble clearing this situation. The comb will have to be destroyed, the tunnelling removed and the comb then rebuilt. Strong colonies rarely show evidence of Galleriasis.

TIP 315: *Deformed wings probably mean a viral infection*

Varroa mites seem to be a significant contributor to the spread of deformed wing virus (DWV), one of 18 viruses known to affect honeybees. DWV is a ribonucleic acid (RNA) virus that was first described in Japan in 1980, and is primarily concentrated in the head and thorax of the bee. RNA is a family of large biological molecules that perform multiple vital roles in gene biology. Only the bees' legs are completely unaffected by DWV. The virus can express itself without Varroa being present, but its effects are much less severe. Possibly, within the hive, Varroa populations are concentrating the virus or even providing a replication incubator for the virus, thus making the effect on bees even more pronounced.

While the deformed wings are the most prominent characteristic, ailing bees will also show shortened, rounded abdomens and paralysis. These virus-infected bees are in serious trouble and will live only about 48 hours. Generally, they are immediately exiled from the colony. The best management method for preventing or clearing this disease is to reduce the Varroa population within the affected colony.

TIP 316: *Know the symptoms of Nosema disease*

Nosema is a troublesome honeybee disease. Apparently, there are two versions of the disease, one caused by *Nosema apis* and the other caused by *Nosema ceranae*. Both diseases affect the bees' digestive systems. *Nosema apis* causes diarrhoea, crawling bees having disjointed wings and no sting reflex. *Nosema ceranae,* though much more vicious within the digestive tract, does not cause the infected bee to faecal spot excessively and crawling behaviour is greatly reduced when compared to *Nosema apis* symptoms. Nosema is not easy to diagnose without microscopic equipment and expertise. The disease is apparently the most prevalent one affecting adult bees.

TIP 317: *Realise that not all dysentery is Nosema*

Dysentery is described as a condition more than a disease. Generally stated, Nosema is a form of dysentery, but other diarrhoea conditions can be caused by winter food stores having contaminants, toxic nectars, fermenting honey, cleaning crushed bee remains within the colony and long confinement during winter months. As with Nosema, brown faecal streaking is a characteristic. Presently, there is no recommended treatment.

ONGOING PREVENTION

TIP 318: *Never totally ignore tracheal mites*

Tracheal mites are invisible mites that live within affected bees' respiratory systems. Their feeding weakens the bees and congests the respiratory system. Until a new beekeeper becomes more experienced with other threats, they can initially ignore tracheal mites. However, for the more experienced keepers who watch their bees closely, these mites should not be totally forgotten. Their pathogenic effect seems to have been reduced to an inconsequential level. To date, no bee disease has ever completely gone away, so completely ignoring a pest that was once considered to be serious seems presumptuous. A few self-assigned experienced beekeepers with expertise in pulling tracheal tubes and checking with a microscope need to stay on the job. Even though tracheal mites are off the radar, they are still out there.

TIP 319: *Never slack on your treatment of Varroa mites*

Varroa mites have taken everything modern bee science has developed, yet they remain the undefeated premier honeybee pest. If your Varroa control programme is working well, you will only see an occasional mite, but even if you don't see them, never forget that they are there and, if given the chance, their populations will once again bloom and overrun the colony. Use monitoring devices like sticky boards and stay in close contact with other beekeepers to get a general impression of the current levels of mite infestations.

TIP 320: *Requeen to control a chalkbrood infestation*

Chalkbrood is a fungal disease caused by *Ascosphaera apis* (see tip 292). For many years, this minor disease was more of a curiosity than a true disease. The condition has become more common. Chalkbrood occurs primarily during summer months and rarely kills a colony, but weakens it to the point that the honey crop is affected.

There is no chemical control recommendation, however, 'Hive Alive' is said to promote intestinal well being from its anti-bacterial fungal and viral properties. Requeening (see chapter 5) is the most common recommendation for heavy infestations. Possibly the requeening process only gives house bees the opportunity to remove infected, mummified pupae from the colony. Often the condition is self-correcting, but the afflicted colony rarely thrives.

TIP 321: *Regularly check pollen moulds and fungi*

Modest amounts of mould and fungal growth within the colony and on stored pollen are considered normal and are not indications of a pathological situation. However, large amounts of mould and pollen do indicate that other problems are present within the colony. A healthy, active colony will not tolerate this extraneous growth. Check the usual colony indicators such as brood levels, queen activity and worker bee populations.

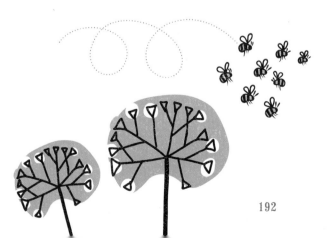

TIP 322: *Not all nectar is beneficial to all pollinators*

The true cause of a plant producing toxic nectar that would sicken or even kill pollinators is undetermined. Some possible reasons for this brew are to:

- encourage specialist pollinators, but not all pollinators,
- protect nectar from microbial degradation and
- protect the nectar resource from nectar robbers.

In addition to toxic nectar, toxic pollens are also produced by some plants for similar reasons. Common plants that seemingly produce toxic nectar are andromeda flowers, oleander, mountain laurel, TiTi (leatherwood) and azaleas. Bees can also become intoxicated from alcohol in fermented nectar or fruit and will exhibit the similar staggering movements made by intoxicated humans.

TIP 323: *Diagnosing pesticide issues is frequently difficult*

Diagnosing pesticide issues is frequently difficult. Harmful bee effects of some earlier classes of insecticides were easy to determine. Thousands and thousands of dead bees would pile up in front of the colony and few adult bees would remain inside. Increasingly, modern pesticide modes of action have become more sophisticated and complex. Gone are the piles of dead bees. Now bees more often die in the field (in reality, many bees also died in the field when exposed to older classes of insecticides), resulting in the mysterious dwindling of the colony's population. Diagnostic confusion results when the beekeeper must include other possible causes such as virus ailments, the concealed effects of Nosema, and the brood output of underperforming queens. Without chemical confirmation, conclusively designating pesticide exposure as the cause of a failing colony will be difficult and subjective.

TIP 324: Treat sparingly

✸ If the proactive beekeeper were to treat for all maladies all the time, ongoing treatments for American foulbrood, Varroa mites (and possibly tracheal mites), Nosema and small hive beetles would be constantly required. Additionally, the beekeeper would have supplemental feedstuffs in place and be watching for Chalkbrood and European foulbrood. Use beekeeping common sense. Monitor your colonies and treat when required.

TIP 325: Know that bee diseases do not affect humans

✸ At this point in our understanding, there is no indication that any bee disease affects humans. Secondary effects of toxic honey may upset human digestive systems, but that is rare. People do not get bee diseases.

TIP 326: Infested American foulbrood colonies can't be saved

✸ If a case of American foulbrood shows up, it is common for a beekeeper to wonder if the colony can be saved. In some instances, it can possibly be pulled back from the brink, but it has been shown time and again not to be worth the effort and the subsequent responsibility of AFB spread. This is not a superorganism that defies control, but it is a very persistent and well-adapted disease of honeybees. Don't tinker around with this pathogen. It can affect all your present colonies and infect your future colonies. Though always disheartening, the surest control for this disease is to destroy all brood combs and, at least, scorch the hive insides. If you have never done this before, read the literature and possibly speak with your local bee inspector from the Department of Agriculture. Aggressively control the disease, but only ask a few trusted friends for advice. Uncontrolled American foulbrood is understandably a troubling issue for other beekeepers.

TIP 327: *Consider how best to deal with Nosema*

Whether or not to treat for Nosema is a consistently troubling issue. Without technical expertise and a microscopic examination, only visible cues can be used to diagnose Nosema infections. Diarrhoea smears on the outside, and in extreme cases on the hive insides, are by far the most obvious clue. But that only diagnoses a dysenteric condition and not necessarily Nosema. In the recent past beekeepers applied fumagillin as the only treatment for Nosema. However fumagillan or Fumil D has since been withdrawn from the market and no longer available in the UK. Its replacement has been Vita Feed Gold which is claimed to reduce the number of Nosema spores and also to generally boost the honeybee's immune system.

TIP 328: *Grasp that bee louse isn't a huge problem*

The bee louse *(Braula coeca)* is an incidental minor pest of honeybee colonies. At this time, there is no known issue with this highly specialised, wingless fly. To the untrained eye, the bee louse can look a bit like Varroa. The primary characteristic of the fly's presence is erratic, thread-like lines that are just beneath honey cappings. They are not found in brood cappings. These are tunnels that the immature fly constructs. They are of little consequence to most beekeepers, but to one who is entering comb honey in honey shows, the tunnels are considered to be comb defects. Honey will seep from these small injuries and give the comb surface a shiny look. *Braula* is found in small numbers in most colonies, and is (thankfully) not considered to be a big pest.

TIP 329: *Stay in touch with your agriculture departments*

As a registered beekeeper you'll be contacted if there's a major outbreak of an exotic disease, or an exotic pest has been found in a port, but it's still worthwhile checking websites of the various agricultural departments of governments from time to time. They have regular updates and new information about beekeeping that are both interesting and accurate. Your local office would be most relevant when it comes to regulations, but they all cover topics that are relevant to beekeeping generally. Beekeeping supplier websites can be an excellent source of new information too, or old information if you're new to beekeeping (see page 288).

TIP 330: *Select beekeepers to ask for advice and assistance*

To grow in both confidence and ability, you just can't beat having an accomplished mentor at your elbow. This type of relationship is excellent for gaining hands-on experience before going solo. Certainly, bee meetings are an excellent place to meet people, but ask extension or state resources for the names of people who teach practical beekeeping. Never forget that, at one time, every beekeeper was a new beekeeper. There is nothing wrong with asking questions and being inexperienced. Later, the beginning beekeeper becomes the experienced one who helps other novices. It's the natural cycle of things.

TIP 331: *Accept that not all colonies will grow to be strong*

🐝 Though much information has been offered here and in other beekeeping resources, always accept the fact that not all colonies will always be great colonies. That is simply not going to happen. Bee colonies are always waxing and waning in size, output and quality. Your job as a beekeeper is to always monitor the struggling colony and help when practical. For your good colonies, your job as a beekeeper is to help them stay good. Bees do not necessarily like us, but it is a fact that bees need us.

TIP 332: *Most importantly of all, appreciate that we can be the bees' greatest friend*

🐝 Beekeepers are the greatest ally and yet the greatest pest with which the colony must contend. We are the bees' greatest friend and greatest enemy. As human interlopers, we are intruders who routinely blow by the best defence the colony can provide. Combs are moved about, bees are crushed and queens are manipulated. Were it not for us, honeybees in most of the world would not have Varroa mites. Were it not for us, Africanised honeybees would not exist. Beekeepers are the primary method of spread for American foulbrood. Yet, because of us – the bee lovers of the world – honeybees have literally been spread over the Earth. Always try to help, never to hurt.

THE BEEKEEPING YEAR

To a greater or lesser extent, there are common tasks that are performed during each beekeeping season. Some of these tasks are necessary, while others are elective. Management systems are constantly evolving. As bee populations have declined, assisting the bees, where possible, has become more important. Supporting the colonies through the seasons is more significant than ever and, hopefully, it will become a task that you take pleasure in doing.

REVIEWING NEEDY COLONIES

TIP 333: *Be mindful that colonies can starve*

The transitional season between winter and spring frequently raises a beekeeper's hopes that colonies will survive, but right at spring's front door, colonies sometimes die due to starvation. A cluster of bees can live all winter on a startlingly small amount of honey – maybe 2.26–3.6kg (5–8lb). But when the brood production starts, the heat generation requirement of the colony will be greatly increased. As a fuel source for that brood nest heat, bees will consume and metabolise dramatic amounts of honey. Spring starvation is not uncommon. You should not become too relaxed until the spring season has truly arrived.

TIP 334: *Heft the colony from the back to estimate weight*

To estimate the survival potential of a colony, beekeepers commonly lightly lift the colony from the backside. Depending on the season and the type of equipment on the colony, an idea can be quickly obtained about the food stores within the colony. If in the autumn, a colony 'heft' feels like deadweight and cannot be lifted – good. If a colony is easily tipped and is obviously lightweight, it will probably have wintering problems. You must decide the future course of action. (See chapter 5.)

TIP 335: *Opt for thick syrup feeding of needy colonies*

When a beekeeper discovers that a colony was not successful in storing the necessary amount of winter food, the most common management response is to feed the colony. Using any number of proven feeding devices, beekeepers routinely feed colonies syrup made from common table sugar to help either a needy colony or a very young colony. Undertaking the successful supplemental feeding of an underweight colony will generally require a warm winter season and a consistent commitment to the feeding process. Thick syrup will be required.

Feeding hungry bees thin syrup, though much easier to make, results in large quantities of water having to be removed by the bees. This is processing work for the bees and results in a high humidity situation within the hive. Water weighs about the same as granulated sugar – about 3.75kg (8.3lb) per 3.8 litres (1 gallon). A thin syrup would be a 1:1 weight ratio and requires about 3.75kg (8.3lb) of table sugar to 3.8 litres (1 gallon) of water. A more appropriate survival sugar feed is a ratio of approximately 7.5kg (16.6lb) of sugar to 3.8 litres (1 gallon) of hot water. Some bee colonies can be kept alive with this emergency system.

TIP 336: *Assess the evidence on dead colonies*

In spite of everything we try, sometimes a colony dies. The reason for the colony death is important. Starved bees will be stuck in cells head first. Watch for evidence of American foulbrood (AFB) or Varroa mite predation (see chapter 7). If possible, check for the presence of a queen. Determine if mice or other pests were inside the colony causing confusion. It is not always a pathogen that kills the wintering colony. Colonies can die with food stores they could not access. To help with future colony management, try to find the true cause.

TIP 337: *Use honey from dead colonies to boost surviving ones*

If a wintering cluster does become isolated from its wintering food stores and starves, what becomes of the remaining honey? The common answer is that the honey is OK for subsequent bee use. Even though it is likely that nothing is wrong with it, do not use it for human consumption. You would need to be sure that no bee disease was the cause – particularly American foulbrood (see tip 297). No doubt something kept the colony from flourishing, but that cause could have been something other than diseases. Poor queen genetics is a common cause. Alternatively, a brutally cold winter may prohibit the winter cluster from relocating nearer its food stores. In general, you can use the extra honey next spring to initiate new colonies, but be cautious.

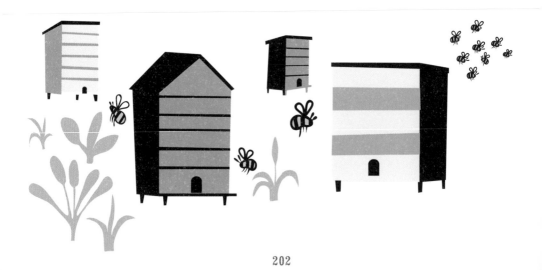

TIP 338: *When weather permits, perform an early inspection*

❀ The presence of a healthy and functional queen is a key component to a healthy and functional colony. Approximately 4°C (40°F) is usually the lowest temperature that a colony should be opened and only then during the early part of the day. Check the honey stores and be certain that there is little colony moisture accumulation. Note the location of the cluster that will be covering the brood nest. If the cluster is clearly up against the inner cover, this colony probably does not have enough stores. Alternatively, if you must peer down between frames of honey to see the cluster, all is well on the food supply front. Only on the calmest days should the brood nest be opened. Most of the time, looking at everything else should be enough to form your opinion.

TIP 339: *Analyse the compost pile in front of the colony*

❀ Experienced beekeepers know that useful information is plainly presented by the bee corpses and colony litter that accumulates at the front of the wintering colony. Ironically, a wintering colony with nothing on the landing board or on the ground in front of the colony is likely already dead. Adult worker bees are dying at fairly regular intervals all winter long. Too many dead workers (hundreds and hundreds) and something is wrong. A suitable number would be more like 10–50 bees or so. Numerous pieces of wax scattered on the landing board and on the ground in front are signs of mice inside the hive. You're stuck: if you open the colony to deal with mice, the wintering cluster will be disrupted, and if you don't get the mice out, then they will disrupt the colony. You're probably better off just to leave things. Animal droppings around the hive indicate badger, sheep or fox activity and these animals may require professional assistance. They don't always do a lot of damage, but neither do they help the wintering colony.

SEASONAL HIVE MANAGEMENT

TIP 340: *Interpret hive-management calendars*

Due to honeybees' remarkable adaptability as a species, they are adaptable and tolerant and can thrive in many different climates. A colony-management calendar is common and useful, but must be tailored to the local conditions. Just like gardening, the colony-management calendar is a historical estimation – a prediction. New beekeepers should explore their use but realise their limitations.

TIP 341: *Clean out your hive every few years*

In general, within a bee nest, spaces of 6mm (¼in) and less are filled with propolis, spaces of 6mm (¼in) to 9.5mm (⅜in) are left open to be used as pathways and any space greater than 9.5mm (⅜in) is filled with combs. Those measurements are mostly correct, but not precisely followed by all bee colonies. Over time, house bees will fill all spaces and cavities, making frames difficult to remove; therefore, the hive must be routinely opened and scraped to rid of propolis and burr comb accumulations. This is not an enjoyable task, but even the finest manufactured hive equipment will clog with propolis and wax if not cleaned every few years. Perform this hive-maintenance job as you can. It is not a regularly scheduled task.

TIP 342: *Expect occasional cold-weather relapses*

❀ Obviously, weather issues are beyond the control of either the bees or the beekeeper. Late-season cold spells are readily tolerated by healthy colonies, but small colonies or colonies with scant honey stores can be in big trouble. In a perfect world, the prudent beekeeper will have held back full supers of honey to provide late-season stores to needy colonies or to provide foodstuffs for newly installed bees. Most beekeepers simply extract the extra honey and hope that all will go well. Feeding these colonies might be helpful in some cases. Sometimes, the weather simply is not helpful. All we can do as beekeepers is make our best guesses and hope for the best.

TIP 343: *Deal with low-producing colonies*

❀ It will be the unusual beekeeper who succeeds in getting all their colonies to essentially be healthy clones of each other, and more than likely, it won't last. In general, the colonies in an apiary will fall into three groups: ❶ need help, ❷ are doing OK and ❸ colonies that are successful. If all the three colonies classified as great colonies are taken and put into a separate apiary, over time, those colonies will subdivide themselves into the three typical groups. Beekeepers should be aware that the health and productivity of colonies waxes and wanes. A lot will depend on the characteristics of the reigning queen. If you feel the current queen is not the lacklustre colony's problem, possibly just adding some brood will pump up the flagging hive. Don't add too much. Monitor for bee diseases in order not to be responsible for spread. But if a smallish colony has not prospered, the queen may be an issue or the nectar flow is passing. Combining this colony with another might be the way to go.

TIP 344: *Evaluate whether to dispose of your queen*

During the first spring inspection, you will begin to form an opinion of the queen's performance. For several reasons, it is not uncommon for some beekeepers to become emotionally attached to a particular queen. The queen is only going to thrive for about a year and then she will begin to fail. There is no demand or traditional use for an old queen. Some possible ones are using her in an observation hive or heading a small teaching colony to be used with children. Normally the queen should be quickly and respectfully killed. She is no longer welcome. If you don't do this deed, the bees will, and they won't be as quick.

TIP 345: *Never forget, requeening every year solves problems*

Many old bee books will recommend requeening every 2–3 years. That recommendation is showing age and is increasingly outdated. Today's colonies should probably be requeened every year (see chapter 5). The reasons for the decline in honeybee queen quality are undocumented and unclear, but complaints abound. There are lots of variables in the requeening process. Is the queen really bad and are your requeening skills good enough to pull this off? If you have kept bees for only a year or two, ask a fellow beekeeper with more experience to help you. Once everything is in balance and you are confident, requeen every year.

TIP 346: *Reverse brood chambers to allow for nest growth*

In the earliest days of using traditional beehives, a single deep was thought to be enough. Queens became more productive and management systems improved, resulting in the average hive needing more space. Nowadays, the core of the average hive is two deep hive bodies. Since bees have a strong tendency to move upwardly onto winter honey stores during winter months, this commonly results in the cluster and the new brood nest being high in the second deep. If left on their own, the bees and brood will develop high near the inner cover. In spring, the colony will consider itself to be crowded and swarming commonly results. Reversing brood boxes is a management procedure that exchanges the positions of the deeps. The reverse must be done during early management procedures. One reversal is usually enough to get the bees to use both brood boxes for the brood nest.

TIP 347: *Clean and maintain bottom boards*

During the reversal process or at any other time the bottom board is at hand, you can give it a cleaning. Sometimes just bouncing the litter off will be enough. Beekeepers commonly scrape the bottom board clear of burr and propolis, but understand that bees will quickly rebuild these burr combs. Scrape the bottom board occasionally, but don't fret over the task.

TIP 348: *Keep hive entrances reduced during winter months*

Mouse invasions can usually be prevented by using colony entrances that are only 9.5mm (³⁄₈in) deep. This shallow entrance can run the entire length of the landing board. Reducing the entrance is primarily done to keep out mice not coldness. Entrances can remain shallow all year round.

TIP 349: *Don't be too concerned by mouldy frames in winter*

If colonies die and leave unprotected honeycombs or if excessive moisture is present, opportunistic moulds will grow on combs and frames. Within reason, these moulds are not a problem. Use these frames again and the bees will clean them next season. These moulds are not pathogenic to humans or bees.

TIP 350: *Replace damaged or worn equipment*

During times when deeps are being reversed and bottom boards are being cleaned, take the opportunity to replace worn or damaged equipment. During the winter season, mice sometimes cause damage, or a frame simply breaks, or there is too much distorted comb. Many times, the aging deep hive body will begin to rot at the corners and will need either repairing or replacing. Making these changes early in the season will be easier as the colony will be smaller and lighter.

TIP 351: *Super before the bees need it*

It is better to provide space before the bees need it rather than after. Bees building up in the early spring will need space for the brood nest to develop. Just a few weeks later, bees will need extra space to process nectar and store newly produced honey. If brood nest space is not provided, either colony growth is limited or swarming will result. Providing extra space after swarm cells are seen will usually not prevent swarming. Once the swarming stimulus has started, it is difficult to change the colony's collective mind. If honey storage space is restricted, a smaller honey crop will be the result. Additionally, the colony will be jam-packed with honey, making the hive difficult to open.

THE SPRING/SUMMER SEASON

TIP 352: *Watch for swarming tendencies at fruit bloom time*

🐝 Every beekeeper, no matter the level of experience, must deal with swarming. Spring swarming is the basic reproductive procedure for all bee colonies and a very strong instinct it is. Eliminating all swarming events is simply not possible. Colonies that winter well and build up nicely are headed by last year's queen and are prime candidates for a big swarm. Letting a swarm issue and you subsequently capturing it is a plan, but not a great one. Anticipate swarming and try to suppress the behaviour as much as possible (see tips 167–182). Expect to fail sometimes.

TIP 353: *If you want to make colony nucleus, do it now*

🐝 While swarming is potentially a problem while the spring pollen and nectar flow is underway, this is the perfect time to make colony nucleus. These divides will have time to build into proper colonies and may be a way to reduce the swarming tendency. Obviously a queen will have to be made available to the new split and simply ordering one is probably the best way to provide a queen for the fledgling colony. But during this time of the year, swarm cells will likely be in some of your colonies and can be used to provide a natural queen to the colony. If natural cells are used, don't skimp on the comb that is cut to move the cells. The tiniest blemish will result in the cell being destroyed. Also, consistently requeening with queen cells will promote the swarming tendency in all your naturally requeened colonies.

TIP 354: *If your choice of hive is the National, move to 10 frames per box after frames are a few years old*

🐝 Even if you regularly scrape and clean your colony, as time passes, the combs will become fuller and tighter. As the first of 11 frames (see chapter 2) is removed, bees are rolled off the comb and some are killed. Though the hive is sized to house 11 frames, using 10 gives more space to remove the first comb and does not roll bees nearly as badly. Commonly, the bees will enlarge the tops of the combs where honey is stored. After combs are drawn, using 10 frames in a 11–frame box is OK.

TIP 355: *Decrease storage space as the spring/early summer nectar flow slows*

🐝 Many, many times new beekeepers state that, 'They had put on frames of foundation, but the bees did not draw it out'. It's an easy mistake to make. Bees will only produce new wax when everything is full – even themselves. Consequently, as the nectar season shows signs of ending, frames containing just foundation should only be used if really necessary. By reducing empty equipment near the end of the season, the bees will fill the existing equipment and complete the honey capping process.

Over super at the beginning of the nectar and under super near the end of the nectar flow. Bees tend to chew unused foundation, but no real harm is done. If the flow is waning, but still ongoing, bees will sometimes 'chimney' or build only on the centre frames of several supers. The frames must be combined into full supers or extra equipment will have to be handled. If an error must be made, go with over supering, but know that the bees will only 'fiddle around' with unneeded super space.

TIP 356: *Remove the spring honey before hot weather comes*

As they do with pollen, house bees mix honey crops within the colony. Speciality crops that have good market value should be taken off as soon as the nectar flows comes to an end. Otherwise, the bees will blend the crop with subsequent crops as they come in. Such naturally blended honey is commonly called 'wildflower' or 'natural' honey. Sometimes it just is not possible, but if it is possible, take honey from the bees before the summer heat arrives. Honey is heavy and sticky. During summer, bee populations are large and most bees are unemployed. It can be hot, heavy work.

TIP 357: *Use specialised equipment to help harvest honey*

There are many devices that are useful in relieving the bees of their honey crop. So many options being available mean that there is no one way that is better than all the others. For only a super or two, a simple bee brush can be used to brush the bees away. Interestingly, the bees don't appreciate being brushed off the combs so suit up. A goose wing is much more gentler and the bees don't seem to be annoyed by them. Many types of 'one-way' devices, called bee escapes, are available for allowing bees to exit the super but prevent them from re-entering. For these devices to work well, the night time temperature should drop enough to entice the bees back towards the brood nest. Chemical repellents are available, too. Blowers, like leaf blowers, can be used to blow bees from supers. They work well, but bees will be flying everywhere. Things will be chaotic.

TIP 358: *Temporarily move and store the crop*

🐝 Provide 'drip boards' beneath a stack of supers and put a hive outer cover on top of the stack of full supers. Never haul open supers on an open trailer. Though it can sit for a long while, process the honey speedily.

TIP 359: *Watch for robbing activities*

🐝 If no nectar source is available, bees will attempt to rob other bee nests. Robber honeybees are nearly maniacal and, once started, robbing behaviour is difficult to stop or to control. Unprotected supers full of honey are tempting targets for honeybees turned robbers. Therefore, supers with bee escapes in place should be structurally sound without alternative entrances. If necessary, tape all cracks with duct tape. Bee blowers, while a highly efficient device for getting bees out of supers, also aerate the area with the sweet smell of honey at a time when no other nectar is available.

TIP 360: In hot weather, stagger supers for ventilation

As the warm season turns to the hot season and as bee populations max out, it can easily become too hot inside the brood nest. Foragers will gather water for cooling the nest, but increasingly, unneeded bees will move out and hang on the outside of the colony. In reality, this is not a harmful event, but do not let it go on too long. Staggering the upper supers about 1.2cm (½in) provides an upper entrance on either end of the super and lets the bees improve the ventilation needs. Guard bees will quickly assume their duty and bees will readily use the entrance.

TIP 361: Be cautious, as summer bees can be hot and cranky

Be prepared for a Jekyll and Hyde experience with a small, growing colony. In the spring time when the colony is smallish and a good spring nectar flow is ongoing, honeybees almost seem like miniature teddy bears. That same colony, at full strength, in late July to August will seem like a completely different colony. The difference is that the colony now has a large population and since the nectar flow is over, most of these forager bees are unemployed. There are more bees in the colony and all of these bees are on 'high alert' watching for robbers bees. In many ways, this is not the same colony that you dealt with last spring. Summer bees can be agitated.

TIP 362: Leave big colonies alone in late summer

Although you have been advised to combine weak colonies or not to make late season nucleus, if you have smaller colonies in hot months, they are much easier and enjoyable to work than the big bruiser colonies. As much as possible, leave the big colonies alone when it is hot.

THE AUTUMN/WINTER SEASON

TIP 363: *Savour the smell of autumn honey*

One beekeeper used the term 'funky' to describe the odour of honey that bees produce in the autumn. You will notice this odour and think your colonies have a serious disease. The honey from autumn flowers does have a distinctive odour but after becoming familiar with it, the odour is not offensive. As seasons pass, the odour becomes a pleasant indicator of a successful autumn honey season. Develop an appreciation for this smell.

TIP 364: *Use autumn honey as winter bee feed*

Consumers accustomed to eating full-flavoured honey (heather, sycamore and dandelion) find milder honeys (borage and oil seed rape) to be so bland as to be syrupy. Even so, the milder honeys are in greater demand for human consumption. Therefore, strongly flavoured honey varieties such as ivy are frequently left on the colony for the bees to use as winter feed. This honey is edible and safe in every way except for the strong odour and flavour. Bees seem to winter well on the crop. Although autumn honey varieties will granulate, they do so slowly.

TIP 365: *Treat for mites, as it is the last chance of the season*

🐝 The beekeeper must develop a personal philosophy concerning mite control and decide whether or not to use chemicals and if so, what chemicals are acceptable. Whatever the decision, this autumn window will be your last chance to do something to suppress the resident mite population within the colony. Attempting to control mites during the winter is a lost cause.

TIP 366: *Make late-season queen decisions*

🐝 Queen-management advice points are nearly as numerous as advice points for Varroa management. The reason is that unless the queen is failing badly or even absent, the decision to leave her or replace her is not a clear one. The recommended action is that colonies be requeened every year (see chapter 5). In fact, most beekeepers will not follow that advice. Queens are normally expensive and can be difficult to get when the need abruptly arises. In early autumn, you can possibly make serious queen changes, but in late autumn – unless you live in a very warm climate – serious queen changes are much less practical. A common solution to a dysfunctional queen problem is to combine the colony with another colony that is not having queen issues. If, perchance, that colony then survives the winter in good form, later next spring that colony can once again be split into two colonies. Queen replacement always requires finesse and confidence from the beekeeper.

TIP 367: *Small hive beetle may arrive on our shores*

The small hive beetle (SHB) is a serious pest in US and many other countries in bee colonies. Like Varroa the small hive beetle will arrive on our shores sooner than later. The SHB is regional; within the region they only infest selected colonies, and within infested colonies, some are harmed more than others. Severe infestations are very unpleasant. The beetles produce slime and destroy combs and equipment. Honey supers awaiting extraction are particularly at risk. There is a chemical controls, but many beekeepers decide to use various models of traps. The SHB is drawn to close places within the hive. SHB traps focus on this behaviour. They are generally more active during warm months, but traps remaining within the colony will not pose a problem. When the beetles are a problem, treat at some time during the season.

TIP 368: *Make sure hives are deadweight heavy in late autumn*

With a combination of luck and skill and with the co-operative efforts of both the bees and the beekeeper, by late autumn the colony should be deadweight heavy. Tilt the hive from the rear. A good colony will be nearly impossible to tilt. Traditional colonies winter in two deeps, but in warmer climates a single brood chamber may be enough to survive the winter. Worried beekeepers frequently leave a honey super on as a hedge against a severe winter. This is no problem for the bees, but if the honey stores are needed, there is a good chance that the brood nest will be initiated in that top super. Again, this is not much of a problem for the beekeeper, but it will take some time to get the brood cycles out of the super and to once again use that piece of equipment as a honey super.

TIP 369: Be prepared to give your best guess on wintering needs

The fact is that some colonies will come up short some years. The beekeeper will be forced to take a guess at the colony's chances for wintering success. Feeding the needy colony is an option, but not a great one. Appropriating honey from a more successful colony and giving it to the weaker unit is another option. Combining colonies is the final solution. Insulating colonies or some other such managerial procedure is not useful. The best solution is to provide space while the autumn flow is in progress.

TIP 370: Combine small colonies now

The number of hives a beekeeper manages is generally a relative number. Colonies can be combined and, the combined colonies can be split again later. Late in the season and as a last management resort, light colonies can be combined with heavier colonies. Don't make mistakes at this late date. If you have any concerns for the other queen, re-cage her for a few days until the uniting process is complete. It would be frustrating to cause the death of another queen. The most common procedure is to put the smaller unit on top of the large hive and temporarily separate the two with a single sheet of newspaper. Punch a hole or two in the paper to entice the bees to chew the paper away. This process will take about two days.

TIP 371: *Remove supers and queen excluders now*

✻ No great harm is done if extra supers are left on the wintering colony, but do not leave queen excluders on the colony through the winter – especially if frames of honey are above the device. If the colony needs the stores and moves to them, the queen will be trapped below the excluder. Naturally she will die there. If you do decide to leave extra empty supers on the wintering colony, reposition the inner cover between the wintering cluster and the empty supers, where it will serve as a partition.

TIP 372: *Keep mice out – all the time*

✻ While there is never a good time to let mice into your bee colonies, now is certainly a prime time to keep mice out. As the season changes and the bees are quiet, mice will begin to move into the colony and build nests in unused corners of the hive. If you put entrance reducers on the hive after the mice are inside, the mice will be trapped inside the hive and will cause havoc all winter. Put mouse guards on before the autumn season advances.

TIP 373: *Don't worry about hive heat during the winter*

✻ As long as they have access to honey stores, healthy bees can readily withstand cold, dry air for an extended period of time. Wintering bees seem to make no conscious effort to heat the inside of the hive; they only heat the inside of the cluster. As the temperature drops to around 14°C (57°F), the bees centralise themselves into a ball of bees and share communal heat. As the temperature continues to drop, internal bees become active in order to generate heat to warm the remainder of the cluster. They must have direct access to honey to produce this heat.

TIP 374: *Provide an upper entrance – in any climate*

🐝 An upper entrance has always been part of the design of the traditional hive. As snow and ice accumulate along the bottom board, the entrance will become blocked, preventing wintering bees from taking cleansing flights on occasional warm winter days. Some inner covers have upper entrances built into their design. Otherwise, the inner cover can be lifted about 6mm (¼ in) to provide a narrow entrance along the top edge. A common practice is to place a match stick under the crown board at the front of the hive for ventilation. Placing it at the back or sides will cause a chimney draught effect. Either way, provide for an upper entrance – in any climate. As warm air rises from the bee cluster and then cools, moisture precipitates, causing the hive to become either damp or outright wet. This 'hive rain' can chill the bees even more as it drips on the cluster. Additionally, as the weather warms a bit, mould growth is increased in damp hives. Should the lower entrance be blocked, the upper entrance allows the moisture-laden air to escape and also allows an upper entrance for cleansing flights.

TIP 375: *In cold climates, forget winter feeding*

🐝 By early winter, the only supplemental feed that can be given to a needy colony is capped honey in frames. There are no feeders that are appropriate for winter feeding in cold climates. In warm climates, feeding may be possible, but even then it will not be an easy procedure. Granulated sugar can be mounded on the inner cover, around the hand hold, for a bee cluster that is high within the colony and low on stores. Bees will rarely attempt to store or process emergency stores and will use the syrup as an immediate food source. Winter feeding is always a desperate situation for the bees.

TIP 376: *Never break the brood nest apart on cold days*

🐝 If the brood nest is opened on a cold day, the overall cluster will surely be seriously damaged. Individual bees become moribund below 5°C (42°F) and though still alive are unable to move. Breaking the cluster causes individual bees to become separated from the cluster and die before they can crawl back to it. Additionally, cluster heating mechanics will be disrupted to the extent that the heating core of the cluster could fail. A very good reason is required to open the colony during this time.

TIP 377: *Locate your colonies in direct sunlight*

🐝 Locating colonies in sunlight has become a general recommendation. Older books may suggest that colonies are in shaded conditions to help with summer heat. It is true that bees may cluster outside the hot hive and even be stressed by the heat, but they are much more stressed having to generate winter heat from stored honey. Being located out of prevailing winds would also be a desirable attribute of a wintering apiary. Colonies can be insulated to help with heat retention, but on the rare warm winter day, that same insulation will prevent the sun's heat from getting into the hive.

TIP 378: *Don't fret about a few dead bees in front of the hive*

Winter bees can live about three months, while warm-weather bees only live a few weeks. Year-round, bees are constantly dying and sooner or later, they are replaced within a healthy colony. In climates where snow is common, dark bees lying on snow are easily visible. New beekeepers may be surprised to see dead bees on the snow and around the hive. To a point, this bee die-off is natural and can be considered to be a good sign if it involves 100, maybe even 200, bees. Dead bees littering the whole area would be a sign of an unhealthy colony that is experiencing a serious die-off. Without snow, these conditions will not be as obvious. Basically nothing can be done to help such failing colonies. Disease treatment times have long passed.

TIP 379: *Expect untreated Nosema infections to show up*

Dysentery diseases like Nosema are not always apparent and treatment is cumbersome. Bees fouling their colony with excrement and excessive faecal spots showing up on nearby surfaces are indications that the bees are not wintering well on the stores they have. Once these symptoms begin to show, nothing can be done to help the bees. In general, if bees can take successful cleansing flights, they might be able to hang on, but this type of colony is not wintering well. At best, it will be in a weakened condition next spring. Stress is a common cause for these diseases beginning to show their effects.

TIP 380: *A honeybee's wintering scheme is still under development*

Honeybees are originally from tropical climates. They are still evolving and perfecting their wintering biology. Even before the days of predaceous mites, winter kills were not uncommon, but currently, these winter deaths are more numerous. The severity of the winter, the size of the cluster, the health of the bees, the colony's wintering genetics and the location of the apiary are all common characteristics that determine how well a colony can survive.

TIP 381: *Give your bees some quiet time during the winter*

After the winter season has begun, little else can (or should) be done until the bees begin to awaken late in the winter. Occasional emergencies arise such as the top blowing off or a tipped colony, but other than these incidental events, the colony's management is on automatic pilot. There is a special task appropriate for this season. This is a reasonably good time for relocating or otherwise moving colonies. Colonies can be relocated within the apiary or they can be moved to new locations. Obviously, the colonies should be handled gently and they should be moved early in the day in order for the colony cluster to recover from the jostling.

TIP 382: *Prepare your programme for next season*

For the most part, the winter season is the time for the beekeeper to regroup and decide the course for the upcoming season. Equipment can either be assembled or repaired during this time. Many major beekeeping meetings are held during this season of the year. Readings and study for specialised tasks like queen production, comb honey production, or mite control are best done during quiet winter months. Enjoy the break.

HONEY USES

The sweet food product, honey, is the primary reason why the honeybee is so popular and respected around the world. At a time when artificial sweeteners are common, honey continues to be the premier natural, wholesome sweetener. Throughout history, honey has been prized as a delectable, natural food. Today, premium honey is in greater demand than ever.

MEASURING THE CROP

TIP 383: *Enjoy the bonus, large honey-crop years*

Nothing is more fundamental to successful beekeeping than honey production. New (and old) beekeepers perpetually feel that the upcoming year will be the big one. Realistically, the perennial hope should be just for an average year. Indeed, as bee populations have become increasingly stressed, the concept of average seems to be changing downwardly. Average crops now seem to be generally smaller than average crops decades ago. But even so, an exceptional year does still come along. Enjoy those bonus years. Big honey crops don't come every year.

TIP 384: *Avoid marginal queens if hoping for big honey crops*

Queen biology and management is a critical but fluid aspect of beekeeping. Within beekeeping, not using marginal queens for honey production is a goal but not a plan. Every good beekeeper wants a good queen to head good colonies that will produce good honey crops. But most queens don't appear to be good for very long – indeed, even if they ever were good. The simple fact is that most beekeepers are trapped into using less-than-good queens on many occasions. Very few beekeepers casually have 'good' queens waiting in reserve. A harsh but truthful recommendation would simply be to requeen every season. This recommendation can be expensive and time-consuming, but in theory, young, healthy queens should always be at the colony's helm.

TIP 385: *Expect small colonies to use much of the honey crop*

🐝 After the main nectar flow is over, if a colony has a small population of workers and if a new queen is successfully installed – as recommended – for the remainder of the season, nearly all the honey that colony produces will be needed to produce workers and will not be stored. The healing colony may grow as quickly as possible and add many new workers, but the honey crop will probably remain scant. This recovering colony would be very suitable for intensive feeding in mid- to late autumn in order to provide it with needed winter stores. In ideal situations, the autumn nectar flow can go a long way towards meeting winter honey stores, but autumn nectar flows are more erratic than spring nectar flows.

TIP 386: *Realise that large colonies frequently mean large swarms*

On one hand, beekeeping is very simple while on the other hand, beekeeping is remarkably complex. If a colony is kept healthy and is managed wisely, and if a productive queen is in place and all is going well, there is good chance that the colony will build up to swarming. Approximately 50% of the colony will leave and hang in the characteristic swarm cluster on a nearby bush – this is known as a 'primary swarm'. As the swarm flies away, so does the honey crop for that season. Swarming can never be totally eliminated. The impulse is too strong. To reduce it as much as possible, the beekeeper should: ❶ try to keep one- or two-year-old queens in the colony and ❷ provide abundant brood and super space before the colony needs it. Many books recommend destroying swarm cells. No major harm done, but this is really just busy work. Watch your good colonies. If you suspect swarming, nucleus them rather than lose them.

TIP 387: *Have empty supers ready for the flow*

It happens to all beekeepers who have the best intentions. Hive equipment, including supers, should be assembled and prepared before it is needed. If producing colonies are under-supered, part of the crop is lost and the available equipment will be jam-packed with honey. This will make it difficult to remove frames and process the crop. Don't crowd the bees.

ASSESSING NEST FOUNDATIONS

TIP 388: *Cull frames not built around reinforced foundation*

New beekeepers with new hives need not worry much about this recommendation, but experienced beekeepers managing older beehives will commonly begin to acquire frames of combs that will need replacing. Misshaped combs and especially combs that are not on reinforced foundation should be replaced. As happens, experienced beekeepers acquire equipment from others as they phase out of beekeeping, but even new equipment is not exactly the same from one manufacturer to another. Consequently due to this slow blending of various styles of hive equipment, rare is the beekeeper whose hives are precisely the same. In fact, combs should only be used for something like 3–5 years before being replaced. It's not worth keeping poor combs because they only make opening the hive more difficult.

TIP 389: *Use plastic 'snap-in' foundation inserts*

A recent innovation has been wax-coated foundation called 'snap-in' foundation inserts. Indeed, that is how the equipment works – it is snapped into new frames designed to accommodate them. The wooden frame married with a foundation insert results in a heavy-duty comb that withstands the occasional hive move and the rigours of extracting. Once the newly outfitted frame is presented to the bees, they will propolise the cracks and reinforce the foundation. The beauty about this frame setup is that it cannot readily be repaired, but increasingly, frames of any ilk are not repaired. (See chapter 2.)

TIP 390: Test out one-piece plastic frames

One-piece plastic frames and foundation are available from bee supply manufacturers. They are lighter than other styles of frames and foundation, but they are fully assembled and ready to use. For the beekeeper on a tight schedule, this is a highly desirable feature. Generally, these one-piece frames are similarly priced to other frame styles, and since no valuable time is required for assembly, it can be argued that they are actually less expensive. This convenience does not come without issues. ❶ The frames are lightweight and will 'rack' from end to end when fully loaded with honey. The resulting small cracks in the capped honey will allow the honey to 'weep' from the combs, giving it a sticky, shiny appearance. ❷ If plastic frames are firmly propolised in place, hive tools will generally slip by the stuck frame top bar. In fact, these jammed frames can be difficult to remove from the hive. ❸ Wax moth damage (see tip 309) to the wax comb is more difficult to repair. Bees may replace combs with large bare spaces. Expediency is their best feature.

TIP 391: Know that full honey supers are really heavy

To simplify operations, beekeepers will commonly use only one size of equipment. Deep hive bodies (24.5cm / 9⅝in deep) are the most efficient to use but they are also the heaviest. A full deep super will weigh just under 45kg (100lb) and honestly, the handholds are very nearly too small for hefting this much weight. To help address the weight issue, other beekeepers use Illinois-depth equipment (17cm / 6⅝in deep), but about one third more hive equipment will be required for each colony. The weight of each super is lowered, but the investment in the hive rises.

TIP 392: Watch for early fruit bloom

✿ A few plants, such as crocus and maple, bloom very early, but the first real flow generally comes from early fruit blooms. Most of this honey and pollen crop is converted by bees. Consequently, apple or pear honey is rarely marketed. Fruit bloom is a clear indication that the spring season is underway and all colony-management programmes should be initiated. Normally, shortly after fruit blooms, the major seasonal flows begins.

TIP 393: Try to be 'super' smart

✿ The natural bee nest is only superficially like the artificial domicile where beekeepers house bees. There is no supering in a natural colony. When a colony completely fills its natural nest space, it either stops foraging or it swarms. When supers are added or removed, beekeepers are actually trying to manipulate bee foraging and hoarding biology. Early in the nectar flow, too much space should be given to the colony. Simply stated, the beekeeper does not exactly know how abundant nectar will be for the season, so should add plenty of supers to allow for more. As the flow begins to seasonally slow, house bees will tend to work the centre frames of all that extra equipment. Beekeepers refer to this behaviour as 'chimneying'. No harm is done, but extra work is required and the bees may chew the foundation. Try to over-super at the beginning of the flow and under-super near the end of the flow. But certainly have additional supers available to the colony.

TIP 394: *Be wary that bees may chew and distort foundation*

🐝 During a nectar dearth, bees may chew and distort beeswax foundation. Possibly bees use this as a wax source in other places within the colony. When this behaviour is observed, remove unneeded supers. If this distorted foundation is used later, beeswax combs can be drawn on this mutilated foundation, but will result in more transitional combs and drone combs that are not perfectly flat. Plastic foundation can suffer more damage. Once the wax is scraped from the plastic, that surface is no longer attractive to the bees for normal comb production.

TIP 395: *Review the various sizes of supers available*

🐝 The average beekeeper will become comfortable with a specific super depth. While beekeepers commonly have strong opinions, there is little science to support the use of one size super over another. The main issue is cost. A shallow super with frames and foundation will very nearly cost what a deep super with frames and foundation costs. A beehive comprised of five deep supers will require about nine shallow supers to provide a similar nest cavity to the colony. Roughly estimated, the colony using shallows (which is impractical and not recommended) would cost 35–40% more than the colony comprised of five deep hive bodies. Super weight is the primary reason for selecting shallower supers.

TIP 396: *Add several supers at once to be more efficient*

🐝 Adding multiple supers at once (sometimes called 'bulk' supering) is usually a better management scheme for a colony as compared to 'progressive' supering where supers are added as needed. Bulk supering requires more equipment at the ready. In progressive supering, hypothetically, a super can be removed, extracted and put back on the colony, requiring less equipment. While not a strong recommendation, bulk supering may have a stronger hoarding stimulus effect than keeping storage space limited.

TIP 397: *If possible, super from the bottom*

🐝 If there is no upper entrance to the hive, the load from a returning forager must be transported all the way up to the top of the colony where it is temporarily stored in empty cells. Putting the empty super just above the brood nest may push the bees to begin work on the empty super quicker than if it is put on top. No doubt some bee labour would be saved, but it gives the beekeeper a significant amount of extra work. Lifting and repositioning those partially filled supers is hard work. Bottom supering probably results in better honey crops but it also adds work for the beekeeper.

REMOVING BEES FROM SUPERS

TIP 398: *Get bees out of the supers*

Bees won't readily leave the honey crop. The honey larder is the colony's only protection against a long, cold season when absolutely no food supplies are available. As the nectar flow ebbs, it will become clear to the beekeeper that surplus honey can be removed. There are several management techniques commonly used to separate bees from their stores without them becoming needlessly rowdy. Brushes, bee escapes, blowers and chemical repellents are all common devices that do the job. Brushing is the simplest honey removal device, but also probably annoys bees the most. Honey removal is chaotic. The task is sticky and heavy and bees will be flying all about the apiary.

TIP 399: *Use fume boards on warm days*

A quieter way to remove honey supers is with chemical repellents. This procedure is better suited for a keeper who has a few years experience. Chemical repellant products are available from beekeeping suppliers and are used with a fume pad. Since repellant vapours accomplish the job, using the fume pads on warm, sunny days will have the best effect. Some of these products have an obnoxious odour that is not harmful to bees or humans, but it really smells bad. The product and the fume boards should be kept away from buildings or other places where the odour will linger. If it smells so bad, why use it? It quickly drives the bees from supers, and one beekeeper with several fume boards can remove a significant number of supers without riling the bees and without the noise of a blower. Additionally, unlike bee escapes, a second trip is not required. Commercial beekeepers use fume boards. The repellant will not drive bees away from brood – only from honey.

TIP 400: *No doubt, a few bees will be in removed supers*

Regardless of what devices are used to drive bees from supers, invariably a few will still be there. It would take hours to get every bee out. But that honey crop is so sweet-smelling to all roaming bees that even if all bees were removed, it is likely that stray bees would be attracted to the supers and to the honey house. Be prepared for some bees to be around the extracting site and for a few bees to still be in the supers.

TIP 401: *Expect disoriented bees as honey is taken*

If the top super is full and if the bees are really packed in a crop, burr combs filled with honey will be broken in the removal process. Confused and disoriented bees will become stuck in oozing honey and immobilised. Many will recover after the episode is over and they are cleaned by their hive mates. Giving these stuck bees aid is nearly impossible. The beekeepers' hand will be sticky. The bee will be sticky and any effort to move or save the bee will likely result in a sting. Just as some bees get left in the supers so some bees get stuck in broken honeycombs.

TIP 402: *Queen excluders are useful to other beekeepers*

Queens will roam the colony searching for appropriate cells for brood production. During the nectar flow, the brood nest is frequently crowded with incoming nectar. Queen excluders are restrictive grids that prevent the queen from roaming anywhere she chooses. Many beekeepers don't use them, feeling that they also restrict fully loaded foragers from getting to the supers above. However, when it comes to removing supers, they really speed things along.

EXTRACTING HONEY

TIP 403: *Extract honey from the supers as soon as possible*

🐝 Honey is a hardy, stable product that can be stored nearly anywhere. But once it is taken from the bees, it should be extracted as soon as possible. Mould growth will start on sticky surfaces, and other insects like ants, earwigs or roaches may be attracted to the sweet food. Don't let unextracted supers sit around for weeks before processing.

TIP 404: *Warm the honeycombs before extracting from supers*

🐝 Honey can withstand a remarkably wide range of temperatures with little effect. Cold honey is thick and pours slowly. Alternatively, honey warmed to about 14–32°C (58–90°F) flows very smoothly and is much easier to filter. Honey supers should be heated to at least room temperature and higher is better, but the room temperature may become too uncomfortable for the beekeeper. Commercial extracting facilities have hot rooms that are heated to approximately 35°C (95°F), but for the smaller operator, heating honey supers can be tricky. Remember that honeycombs get soft at about 38°C (100°F) and will melt at about 63°C (145°F). Don't try putting space heaters near supers to warm them up. However, if honey supers are removed on a cool day, they will need to be warmed to improve the extracting process.

TIP 405: *Only extract capped honeycombs*

🐝 If the nectar flow is abruptly cut short by weather or harvesting, there will simply not be enough new wax being produced to finish capping the crop. House bees will not re-eat processed and stored honey to metabolise it into wax. It will remain uncapped until another flow comes along that will allow the natural production of new wax. Honey refractometers are used to measure the moisture content of honey. Ideally, it should be about 17% moisture in order to both store well and be pourable. If possible, blend the uncapped frames with other honey that has been capped. Every honey crop will have a few frames that are not fully capped.

TIP 406: *Use heated knives for uncapping honeycombs*

🐝 Heavy, serrated-edged knives are commonly used to cut the wax cappings on honeycombs, with performance improved if the blades are heated in hot water beforehand. Alternating two knives is a solution. Cold honey quickly draws the heat from a knife. Electrically heated uncapping knives are available from suppliers and make uncapping life easier. In larger operations, automatic uncapping devices use two serrated knives that move in opposite directions to 'saw' through the comb cappings. These 'back-and-forth' knives are heated with water or steam. Heated knives always work better but be careful on the temperature, as the honey can become caramelised and the wax can give off a burnt aroma.

TIP 407: *Processing honey is a sticky job*

🐝 When honey spills and dribbles arise, they can readily be cleaned up with warm water. Getting a high-quality honey crop is thrilling but, invariably, processing honey requires frequent hand washing and cleaning of necessary processing equipment.

TIP 408: *Either borrow or buy a honey extractor*

A honey extractor uses centrifugal force to throw honey from uncapped honeycombs. The devices have been around for many years and are invaluable to beekeepers; therefore both new and old ones are coveted. Modern extractors are made either of food-grade plastic or stainless steel. Lead-free solder is used to seal metal joints. Plastic units are lighter and cheaper. They tend to be used by beekeepers with only a few supers to extract. Older extractors are commonly heavy-duty contraptions with solid spinning mechanisms. They seem to last forever, but they nearly always have galvanised tanks. If this old equipment is to be given new life, the tank should be thoroughly coated with a clear epoxy coating.

TIP 409: *Prime honey filters and strainers with filtered honey*

Honey straight from the extractor will have many pieces of beeswax and propolis particles in the unfiltered honey. There are many designs of filters and strainers for removing these natural contaminants. Check with bee supply companies to select the best one for you. Most involve the use of a nylon strainer. These filter bags work much better if they are within a canister that can be filled with filtered honey. As unfiltered honey is added to the submerged strainer, contaminants will stay in suspension. If the filter bag is allowed to run dry, within just a few cycles the contaminants will stick to the bag wall, clogging it and making it ineffectual as a filter.

Honey extracting cycle

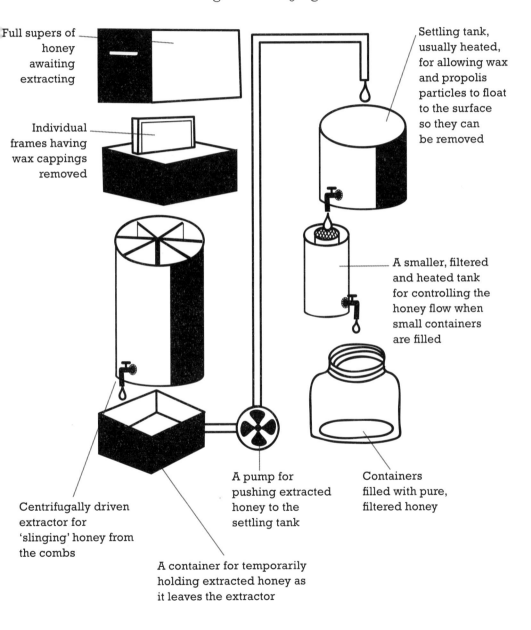

Full supers of honey awaiting extracting

Individual frames having wax cappings removed

Settling tank, usually heated, for allowing wax and propolis particles to float to the surface so they can be removed

A smaller, filtered and heated tank for controlling the honey flow when small containers are filled

Containers filled with pure, filtered honey

A pump for pushing extracted honey to the settling tank

Centrifugally driven extractor for 'slinging' honey from the combs

A container for temporarily holding extracted honey as it leaves the extractor

TIP 410: *Consider buying an electric honey pump*

A honey pump is probably the first piece of equipment that is added to the extracting line after purchasing an extractor and uncapping device. Honey weighs about 534kg (12lb) per litre (3.8 litres); therefore, a 19-litre (5-gallon) bucket of honey weighs 27kg (60lb). Rather than capturing draining honey in buckets, a pump can be used to push the extracted and filtered honey to the settling tank. This pump reduces beekeeper labour and honey spills. These pumps do come at a price and will require the extracting line to be plumbed with food-grade PVC or vinyl tubing. Also, small pumps tend to become hot during use.

TIP 411: *Watch that honey line – it's under pressure*

Clear vinyl tubing is used as a flexible tube between the honey pump and the filter. As the filter bag becomes increasingly clogged, the internal pressure on the vinyl tubing increases. Normally the honey has been warmed so the supply tube is also warm. If the filter should become totally clogged, the pump will continue to push honey into the lines which will cause the line to dramatically balloon as it becomes overloaded. If not shut down, this can result in a large mess.

TIP 412: *Couple rather than glue polyvinyl chloride tubing*

Polyvinyl chloride (PVC) piping should have threaded unions at strategic locations so the plumbing can be disassembled for cleaning, maintenance and storage. The normal small-scale extracting plumbing is a combination of PVC piping and flexible vinyl tubing. If honey is allowed to sit in these lines, it will frequently granulate, causing a clog the next time the system is used. It would be impractical to attempt to flush with water. All the honey in the line would be wasted and water not drained would either dilute the honey or allow mould growth to occur. Disassemble the piping and thoroughly clean and dry it after the extracting season is over.

TIP 413: *Use gentle heat on the settling tank to speed honey clarification*

After honey has been extracted and strained, it is moved – either by bucket, gravity flow or electric honey pumps – to the settling tank, which usually serves as a bottling tank in smaller operations. In the settling tank, small contaminant particles and air bubbles will float to the top, where this layer of natural detritus can be skimmed off. If the tank is gently heated to about 32°C (90°F), contaminants will float to the top much more quickly. More expensive tanks are water-jacketed and are designed to heat the honey to a controlled temperature.

Additionally, if the settling tank doubles as a settling/bottling tank, the honey will flow much faster from the tank, and if heated honey is being bottled, the piston valve will work more smoothly. Electric water pipe heating cable can be wrapped around the unheated tank to supply heat. Heating the settling tank is not a requirement, but it helps.

TIP 414: *Be careful when trying to process 'high moisture' honey*

There should be a good reason why honey with high moisture content has not been correctly processed by the bees. The safest procedure is to give the crop back to the bees and let them finish it. But possibly the season was truncated by weather. Maybe the colonies were abruptly moved from a pollination site. A sudden end to the nectar flow can happen. The beekeeper has only two honey drying options: ❶ an improvised device from web-based plans or ❷ a small commercial unit that may cost a lot of money. The improvised units will probably consist of a small heater and a blower that passively removes moisture from the uncapped cells. It is important that the heat is gentle and low. Too much heat will soften combs. Drying honey with high-moisture content down to the requisite 18.6% level should be a process that is rarely done. If at all possible, let the bees worry about it.

TIP 415: *Check labelling regulations*

It is up to local beekeepers, if selling honey, to be sure that all legislation and food-handling guidelines are followed. The information can be found via DEFRA as well as the Food Standards Agency (see page 288). Local associations such as the British Beekeepers Association website have good online information and should be able to assist. The EU have updated the honey regulations since 2005. Further advice on the production, processing, distribution, retail and labelling of food stuffs can be also be found on the Food Standards Agency website. It would be highly advisable when selling honey or food stuffs to do your homework this could save you money and time. It would be also advisable to be aware of the 2007 amendment on infant botulism: a voluntary labelling code stating that "honey should not be given to infants under 12 months of age"; a precautionary measure against possible infant botulism. Most beekeepers add this to their back label.

TIP 416: *Always watch for bee parts that escaped the filter*

🐝 In the dark hive, bees are intimately associated with the production and processing of honey. More often than not, the public should be quietly protected from some of the details of the bees' honey production procedure. Therefore, it is critical that bee body parts not be allowed to turn up in honey that is sold (or given) to non-beekeeping consumers.

TIP 417: *Consider a temporary extracting setup*

🐝 The goal of many beekeepers whose operation is growing is to construct a dedicated honey house. In fact, plans abound that describe honey house features such as flooring, door widths, lighting, wall coverings, and plumbing and drainage. Other beekeepers elect to keep all the extracting line and related equipment portable. As an example, one beekeeper sets up his extracting line in his garage, where, for about 5–6 weeks he processes several buckets of honey. Once finished, he breaks it all down and the space reverts to a garage. In this way, his income from his bees is not tied up in buildings and related maintenance costs. This setup is not for everyone, but it works well for some.

COMMERCIAL PREPARATION

TIP 418: *Diversify into more than honey production*

🐝 There are many variations that the creative commercial beekeeper can explore. Naturally, large-scale honey production and pollination services are the hallmark of commercial beekeeping, but providing equipment and package bees is always in demand by both new and nearby beekeepers. Other speciality beekeepers may produce comb honey. The nectar sources in the area would affect this decision. Creamed honey (honey sold with a texture like butter) is popular as a breakfast and sandwich spread, plus it can be flavoured with spices like cinnamon. One of the most specialised aspects of beekeeping is producing honeybee queens. The demand for queens is great – especially high-quality local queens. Though real profit can be realised from queen production, it requires a rigorous seasonal schedule. Research all beekeeping specialities.

TIP 419: *Price your product fairly*

🐝 Research the price of honey on the shelves in various local stores and roadside markets. Bee magazines publish honey prices by region. Design a catchy label. Since the buying public may not be comfortable with bees, keep bees on the label to a minimum. Produce a clean, clear product. Flaunt the attributes of your honey. Set up at fairs and markets. Explain the granulation issues. Producing honey is not always an easy task. Ask a fair price for your product.

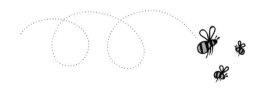

TIP 420: Develop vendor displays

🐝 Vendor display technology is presently visually appealing and easy to set up. At £200–400, folding tabletop displays are reasonably affordable and present a positive image to the public. In many models, the company name is presented in a banner across the top of the suitcase-sized display. The display will need to be near human traffic flow to be clearly visible. Pull-up banners are more expensive, but are taller and more visible from a greater distance. If you will be presenting your company at an indoor session, these displays are appropriate. If you will be outdoors, maybe at craft and garden shows, these displays will not be necessary.

TIP 421: Create a web presence

🐝 In recent years, a diversified web presence has become nearly mandatory. Unless the beekeeper is already adept at webpage construction and maintenance, outside professional help may be needed. Social media sites such as Facebook and Twitter are fundamental ways to get your company's message out to the public. These social media systems are both a help and a hindrance. They require money to set up and constant maintenance. People who write to you will expect an instant response. Indeed, a type of beekeeping is developing where little is actually done in an apiary, but information and vicarious experiences are discussed in detail. Video systems such as YouTube are invaluable tools for beekeepers to promote both themselves and their bee operations.

TIP 422: *If the operation expands, a range of skills is necessary*

Beekeeping is not only about keeping bees. As the operation grows, ancillary skills and abilities will need to grow, too. Installing the plumbing for honey sumps with food-grade plastic tubing, installing an electric honey pump, repairing and adjusting automatic uncapping systems, arranging bottling tanks and maintaining honey filtering systems are all examples of the diversified skills that are needed to help a young beekeeping operation become a commercial operation.

TIP 423: *Do your best to deal with any customer complaints*

Increasingly, as the developing beekeeper is required to deal with the public, issues will occasionally arise that require patience and understanding. It's not always easy. Since honey commonly granulates, the occasional customer may express concern about this product. You may also hear 'This honey confectionary seems stale to me'. A major area of complaint is with queens and newly purchased nucleus. While the bees may be healthy when they leave the producer, they sometimes die in transit.

Alternatively, a beekeeper may be inexperienced in queen introduction and the queen is killed in the process. The producer must decide how to keep the customer happy but at the same time continue to make enough profit to continue in business. Generally, most customers are satisfied, but be prepared when they aren't.

TIP 424: Develop contacts and sources for product

A valuable aspect of a commercial beekeeping operation is the source list for honey and products that evolves over time. Availability of varietal honey producers is an example of a valued contact. A recommendation and introduction are typically required before business is transacted. Methods of payment and an estimation of future purchases are required. Honey drum exchanges are often necessary. This procedure reflects a limited supply situation. These sources and relationships take time to mature.

TIP 425: Uphold common business principles

Many beekeepers are good beekeepers but not particularly good business people. Tax laws must be understood and bank loans managed. Loans and monetary support will depend on your personal credit history. Most successful commercial bee operations also have a source of competent business skills. If you don't have such contacts and skills, you must develop them before proceeding very far in the development of a commercial operation.

SELLING HONEY

TIP 426: *Be clever with your marketing label*

🐝 It is probably not a great idea to put an anatomically correct bee on your honey label. It is due to the bees' diligent work that the product exists, but in the UK, the public does not eagerly eat insects or insect products. Take note of the artwork on some of the labels used by large honey producers. Generally, the company name or appealing flowers are featured. Though, if real photographs of flowers are used, as opposed to clipart, the honey in the jar must be from this flower.

TIP 427: *Use a back label to explain honey granulation*

🐝 Untold amounts of honey have been discarded because it has granulated. Granulation is a reversible natural process. The honey can be eaten or used in cooking in the regular amounts. However, granulated honey will have a gritty texture that may not be pleasant. Honey granulation is an indication that the honey has not been highly processed, but it certainly does look as though it has 'gone bad'. Some of the sugar molecules have coalesced into a crystalline latticework. Granulated honey can be driven back into solution by applying heat of approximately 35–40°C (95–105°F), or drop in a hot water bath. Loosen the lid and shake the honey occasionally. Afterwards, let it cool and replace the lid. Within a few months, it will again granulate. Brief instructions on the back of the jar will reassure customers.

TIP 428: *Consider plastic rather than glass containers*

Plastic containers have replaced the glass containers that were previously used. Though glass still gives a clean, clear view of the product, freight charges on empty glass have increased. Additionally, a glass jar will shatter if dropped or can break inside the case during shipping. Plastic costs are tolerable and the containers are more forgiving than glass containers. Most beekeepers use a mix of containers. Glass is used for 0.45–0.9kg (1–2lb) jars, while heavier containers and squeezy bottles are made of plastic. Overall, plastic is easier to use for honey containers.

TIP 429: *Do not sell your product cheaply. Be realistic.*

In general, honey sells for about £5–10 per pound. Beekeepers are known for being able to work with the bees to get a honey crop, but often struggle when trying to establish customers and a marketing base. The selling price will vary from year to year depending on reserves and crop production. Local honey sold to local people normally brings an above average price. Honey from known sources (blackberry, clover, heather and lime) are also in demand. Much of the honey produced comes from a mixture of plants and is called 'wildflower'. This honey might not bring as much. Visit groceries and craft or market shows to research prices. Go online or visit the British Beekeeping Association website to get an idea of national selling prices.

TIP 430: *Remember, honey is a premium food*

Honey properly handled and bottled is a high-quality food product. Some producers give products to friends and as prizes at meetings. Though a valued product, it is a rugged commodity that can withstand long storage and has multiple uses. A nice jar of honey is truly an achievement.

PRODUCING COMB HONEY

TIP 431: *Make comb honey and extracted honey in different hives*

Extracted honey is generally the easiest honey product to produce. Through the years, efficient and affordable extracting equipment has evolved. Combs and cappings don't have to be perfect and extracting is reasonably easy. Comb honey requires very lightweight edible foundation or virgin wax and specialised comb honey frames. These frames are commonly not liked by the house bees. If they have a choice between a common extracting super, used for producing liquid honey, and a specialised comb honey super, used for producing honey in the combs, they will nearly always choose the extracted super. When producing comb honey, don't tempt your bees by including other types of honey supers.

TIP 432: *Develop a market for your comb honey*

Several decades ago, comb honey was a common product from the beehive, but it is much less so today. Since there is no easy way to adulterate it, comb honey has always been considered the purest of the honey crops. Usually the best combs are used to make this specialised crop. If the comb honey is not eaten within a few weeks, some crystallization may occur, but it is still perfectly usable. But the new comb honey consumer may have other more delicate questions about how to eat comb honey. What should be done with the wax? There are two obvious options: ❶ politely remove the chewed wax in a napkin or ❷ swallow it. Human digestive systems have little effect on beeswax and it is simply voided. Wax melts at about 63°C (145°F) so it would be softened at 37°C (98.6°F). Most proficient comb honey eaters use both options at once. Either way, time and educational energy are needed to teach customers how to eat comb honey.

TIP 433: *Make chunk honey products*

🐝 When producing comb honey, some of the combs are perfectly appealing. A common product that uses these perfect pieces is called chunk honey and is a readily accepted product. Wide-mouthed jars designed as a container for this honey product are lightly filled with a chunk of comb honey. The empty spaces around the combs are filled with extracted honey. The comb should be 50% of the product. The honey in the combs should be the same as the liquid honey, preferably light in colour, in order to better show the comb product. The product is a blend of comb honey and extracted honey and is spooned out gleefully.

TIP 434: *Expect comb honey crops to be more erratic*

🐝 Another reason that comb honey is more expensive and not always readily available is that good comb honey nectar flows do not necessarily come every season. New wax must be secreted and formed into new honeycombs. Then that fragile new comb must be filled with light honey that is all from the same source. Big nectar sources like clover or orange blossom are common comb honey crops.

TIP 435: *Granulated comb honey cannot be reliquefied*

🐝 To prevent granulation, comb honey should be frozen until sold. If it does granulate, about the only thing that can be done is to give it back to the bees and let them sort it out. They will use it as carbohydrate food, but they will not be able to heat granulated honey hot enough to reliquefy it. Initially, quite a few of the honey granules will be dropped to the bottom board, but over time the house bees clean it up and the product is reused.

PRODUCING CREAMED HONEY

TIP 436: Try producing creamed honey

❀ Creamed honey is simply granulated honey with very small crystal sizes. Therefore, it does not have the gritty, sandy texture that naturally granulated honey usually has. To produce the product, raw honey is first heated to kill any yeast that is commonly found in the honey. After cooling, about 10% of previously creamed honey is added to 90% of the extracted honey to provide seed crystals of the desired size and the mixture is poured into shallow, wide-mouth plastic individual containers. The blended honey is then held at 14°C (57°F) for about a week. If creamed honey seed is not available, a seed batch can be made by crushing the crystals in naturally granulated honey. Creamed honey sometimes has cinnamon, blueberries or strawberries added to the mix to produce flavoured creamed honey.

TIP 437: Know what the honey purity regulations entail

❀ The Food Standards Agency for honey regulations mentions that specified honey products should be a natural product with no less than 60g/100g fructose and glucose sugars (sum of both) for blossom honey. Sucrose content in general should be no more than 5g/100g, with moisture content in general not more than 20% for Heather (*calluna*) and bakers honey in general not more than 23%. Bakers honey from heather should be no more than 25%. Also, if anything is added to the honey though, like fruit or an infusion, those ingredients must be listed and the front label must accurately reflect the addition, like say 'Creamed Honey with Cinnamon'.

TIP 438: *Consider mead making for product specialization*

🐝 Some mead (honey wine) is produced to keep some of the honey's sweetness to the extent that it is considered to be a dessert wine. Drier meads and sparkling meads are also available. Some grape wines may have honey added after fermentation as a sweetener and flavouring. In many instances, the mead maker is not a committed beekeeper or may be not a beekeeper at all. Making quality mead products requires significant time allocations and dedication to strict fermenting techniques. Ironically, the earliest meads – dating back about 2000 years in China – were probably no more than honey that was poured into boiled water.

TIP 439: *Use honey to make home-brew beer*

🐝 No doubt honeybeer has been around for many years, too, but its history is not nearly as rich as that of mead. Even so, honeybeer made commercially and by home brewers is very popular. Honey is a logical choice for beer-making recipes. It imparts colony aroma and flavour to the final product. The problem is that boiling honey to kill any naturally occurring yeast also boils off some of the flavourful agents and destroys the enzymatic system that allows fermentation to proceed. Generally, honey is heated, but not to boiling. Many recipes and procedures are available.

TIP 440: *Use lighter honeys for food sweeteners and fermented drinks*

🐝 Honey flavour and variety is strictly a matter of personal preference. If an individual spent his or her early years eating dark honey on cereal, light honey may taste too bland. Since they don't add a strong flavour or aroma, lighter, milder honeys are used as sweeteners and to make mead or honeybeer. If a stronger flavour is desired, use darker honey.

CONSUMING HONEY AT HOME

TIP 441: *Use honey any way you like*

🐝 Honey is a universal sweetener. For food uses it can be eaten directly, it can be mixed with drinks like tea or coffee, it can be fermented into specialty beverages, or it can be used in cooking. Types and flavours of honey vary in any country and from season to season. As with coffees or teas, individuals develop a taste for honey from their region and in a form to which they have grown accustomed. There is no wrong way to eat honey.

TIP 442: *Eat comb honey to enjoy the full flavour of honey*

🐝 Some beekeepers are keen to provide more instruction about producing the higher-value honeys that are sometimes called artisan honeys. Honey comes in many flavours and colours and is harvested in careful ways to preserve its truest essence. To truly enjoy the bouquet, honey should be eaten in the comb. Purists will argue, with merit, that the extracting process that airs the honey in the extractor and then the low heat needed to filter and process liquid honey can reduce the full bouquet of its tastes and aromas. This is a personal choice. For whatever reason, if you are interested in eating honey as the bees produced it, eat comb honey.

TIP 443: *It is possible to eat too much honey*

🐝 In reasonable amounts, honey is a wholesome food. As it is a natural sugar, if consumed to excess, honey can be a problem for diabetics and can cause dental problems and contribute to weight problems.

TIP 444: *Freeze honey for later use*

🐝 Honey is commonly cheaper if bought in larger sizes such as 0.9kg (2lb) or 2.26kg (5lb) containers. Serious honey users will buy it in 19-litre (5-gallon) pails. Most types of honey will granulate over time and while it can be reliquefied, turning it back into a liquid is a tiresome process. Honey can be frozen. This will stop the granulation process, but a personal technique will have to be worked out for handling a large storage container. Frozen honey will not pour. As an alternative, several small serving containers could be filled at one time and a single container removed and thawed as needed. Honey is stable in a closed container in the cabinet, but do consider freezing it.

TIP 445: *Explore the non-food uses of honey*

🐝 Honey is primarily thought of as sweet human food, but many uses exist for honey other than eating it. It has abundant uses in facial and hair-care products. Salves and balms may have a honey component in the mixture. In both personal care products and when used in cooking, honey absorbs moisture from the air; therefore, skin or food dries out more slowly. Honey also produces hydrogen peroxide and as such has been used for many years for skin injuries and for burns. Honey is reputed to have many beneficial attributes that are not always clinically proven. For instance, decide for yourself if honey helps your digestive system or helps you sleep.

TIP 446: *Don't think you can store honey indefinitely*

🐝 Honey is reputed to be a long-lasting, clean product that can seemingly sit on the countertop for months without harm. While this is largely true, there is a rational limit to where and for how long honey can be stored. Honey is widely reported to have been found in ancient Egyptian tombs. But that honey still being edible could be pushing the point a bit far. Honey that has been stored for 20–30 years becomes very dark and thick and certainly none of the delicate floral nuances remain, but it does keep some of its sweetness. Honey also becomes very thick as moisture evaporates. Honey below 17% moisture will essentially not pour. For whatever reason, if honey is going to be kept long term, glass containers with coated metal lids should be used. Honey is amazingly stable and has evolved to be stored by bees in a cavity in a tree, but it may be a long shot to think that honey stored for many years can still be tasty and enjoyable. Use honey within a year or so.

TIP 447: *Experiment with sugar replacement recipes first*

🐝 When cooking with honey, experienced cooks know that the relationship between recipe ingredients and honey varies. A general rule can be given for directly substituting honey for 50% of the sugar, but with testing and fine tuning in some recipes all of the sugar can be substituted with honey. Honey is acidic. Neutralise the acid by adding ½ teaspoon of baking soda for each ¼ cup of honey. Since honey is a thick liquid, it blends easily with other recipe ingredients. Experienced cooks have recommended dipping the measuring spoon or cup in cooking oil in order for the honey to drain quickly and cleanly. The flavour of the honey can affect the texture and the aroma of the cooked dish. Plus honey is generally a bit sweeter than the table sugar it is replacing. The process of using honey rather than sugar in a recipe is not an exact procedure. Test the recipe before going public with the dish.

TIP 448: *When cooking with honey, lower oven temperature*

An important characteristic of cooking with honey is that the dish will brown more easily than recipes containing only sugar. A common recommendation is to lower the oven temperature by about 4°C (25°F). Since honey is about 18.6% water per cup of honey used, other liquids such as water or milk should be reduced by about ¼ cup. These are estimates, so do test the recipe before going public with the dish.

TIP 449: *Baking with honey improves breads and cakes*

The chemical configuration of honey forces it to absorb moisture from the air. It seems improbable, but honey does keep baked goods moist a few days longer than recipes just using table sugar – especially if the honey source is from trees, holding more fructose sugars. Additionally, the texture is made firmer and crisper and less crumbly.

TIP 450: *Explore beverage recipes that use honey*

There are many, many concoctions and mixtures that use honey as an ingredient. One that is very simple and easy to mix is simply honey and mineral water. Add about a teaspoon or two to a glass of natural mineral water at room temperature. Purists require bottled mineral water and not tap water. It makes a simple nutritious drink. Tea and coffee can readily be sweetened with honey. Hot spiced tea can be made by putting ¼ cup of honey in four cups of brewed tea. Add maybe four 7.5cm (3in) cinnamon sticks and about four whole cloves.

HONEYBEE BY-PRODUCTS

While honey is clearly the best-known and most abundant hive product, it is not the only one. Wax, propolis, pollen and bee venom are other hive products that have specified uses and groups of dedicated devotees who routinely use them. Though these extras have speciality uses outside of beekeeping, they bring diversity to the beekeeping craft.

WHAT TO DO WITH RENDERED WAX

TIP 451: *Remember that beeswax is flammable*

🐝 Beeswax is highly flammable. The smaller-scale beekeeper is most likely to start a fire while rendering wax in small boilers directly over an open flame. The best way is slower but much safer, and that is to use a double boiler. Beeswax will flash at about 204°C (400°F). If a small fire does break out, turn off the power source before tackling the fire, using a dry chemical powder extinguisher for a serious fire, fog or foam or fire blanket, but do not use water.

TIP 452: *Separate new cappings wax from old comb and burr comb wax*

🐝 Comb is secreted as whitish beeswax scales and is used to build new white combs. As the comb is used and is stained by pollen, beeswax and larval faecal matter, the comb darkens as the years pass until it becomes jet black. Virgin wax used to cap honey is pristine and will bring the highest wax price or will produce vibrant yellow candles. Wax from dark combs – especially in the brood comb areas – will result in a dark olive-brown wax that has industry uses, but is less valued. Cappings wax should be rendered separately from wax rendered from broken combs or burr combs.

TIP 453: *You won't get much wax from old, dark brood combs*

🐝 Cappings in honey storage areas are lighter both in weight and colour than cappings in the brood nest area. Additionally, the cappings used to cap honey are not of the same composition as the cappings on brood cells. Propolis is added to brood capping, giving them a browner colour and a heavier texture than the pure wax cappings that cover honey-filled cells.

Each time a bee completes a life cycle in brood cells, it lines the walls with a silken cocoon. At a development stage within the bee's life, it defecates within the comb and that faecal residue will stain the brood combs. Worker bees clean and polish these combs in preparation for the next occupant. Over time, these fibrous cocoon residues build up and, along with propolis added to reinforce the lip of the comb, will thicken the cell walls.

After many years, these stout black cells will become smaller and will produce a worker bee that is slightly smaller than workers from newer cells, but few combs last as long as that. When these combs are rendered, the propolis and residual cocoons will make rendering the bit of wax within original cell walls difficult to remove. For best results, a wax press should be used. For most beekeepers, this will not be worth the dark wax result.

TIP 454: *Only use stainless steel pans for rendering wax*

🐝 Any containers or utensils that are used to render beeswax will probably never have any other use other than melting beeswax combs. Large food cans or clean paint cans can be filled with comb pieces and put in a larger and deeper water-filled pan. Water should come about half way up the side of the container. Various pans and devices are available from bee supply sources, but some beekeepers have had good luck using an old crock pot to slowly and indirectly melt the comb pieces.

TIP 455: *Use a solar wax melter to keep your apiary clean*

Invariably, small pieces of wax accumulate as a colony is manipulated. If dropped, these pieces stick to the bottom of shoes and more importantly can spread disease or initiate robbing behaviour. Certainly, these pieces can be saved and rendered as time permits. Alternatively, a solar wax melter can be used to process these incidental pieces as they are removed from colonies. Plans are available on the web for home-constructed units, or devices can be purchased from commercial outlets.

Essentially a solar wax melter is nothing more than a glass-covered box painted black on the outside and white on the inside. The box is angled so it sits higher on one end. Inside, a metal pan holds the wax scrapings. Though one layer will work, the box heats better if covered by two layers of glass separated by 1.2cm (½in) of air space. More complex units are on a pivoting axis so the unit can be turned during the day to continually face the sun. Once in operation, this wax-melting gadget is free and simple to operate, but the high temperatures that develop inside the melter will bleach the wax nearly white. Additionally, much of the wax will remain in the residue that stays (slumgum), so this device is not highly efficient. They are, however, still considered to be a useful piece of equipment by many beekeepers.

TIP 456: *Have slumgum rendered commercially*

❀ The residual remains of the wax-rendering process, colourfully named 'slumgum', are not very appealing. If you can get by the initial negative perception, this material is stable and without offensive odour. Wax moths have little interest in it, though curious bees will buzz about it when no other food sources are available. As wax is rendered, a dark layer of cocoons and propolis will accumulate along the bottom of the wax cake.

The bottom and sides of metal wax melters will occasionally need to be scraped free of this residue product. It is a low-value product, but stores easily. Commercial wax companies will buy slumgum and, under high heat and pressure, will extract the otherwise unextractable wax from the residue. Commercial operations generate large quantities of this stuff, so the rendering effort is worth the cost. Beekeeping groups may communally collect this residual product for rendering. For the smaller beekeepers, no doubt other uses should be found (see tip 457). Regardless, don't throw it away.

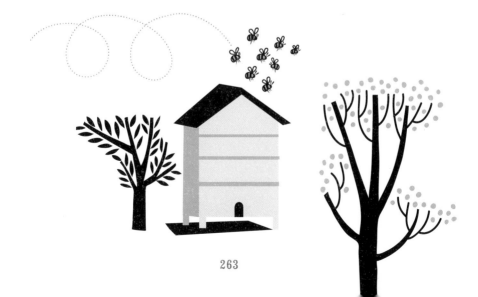

TIP 457: *Find other uses for slumgum*

In general, slumgum is an unloved, frequently under appreciated hive product. Even after professional rendering, it still has a high enough beeswax content to make it readily flammable. This product can be melted and poured over Hessian sacking or kindling and used as an easy fire starter, or it can be used in chunk form as a firestarter. Scout bees have an interest in sun-warmed slumgum, so lumps of it can be used in swarm traps or it can be melted and brushed onto the swarm trap walls. Even if the wax content is low, it has many of the same uses as rendered beeswax. It can be used to lubricate screws or nails and will serve as a simple polish on tool handles. Some apitherapists – individuals who use bees and bee products to address human ailments – apparently find use for slumgum in some of their procedures. The final use is just to bury it as fertiliser and compost. With thought, no doubt other uses could be found for this unloved hive product, but at this time, most slumgum is either burned or buried.

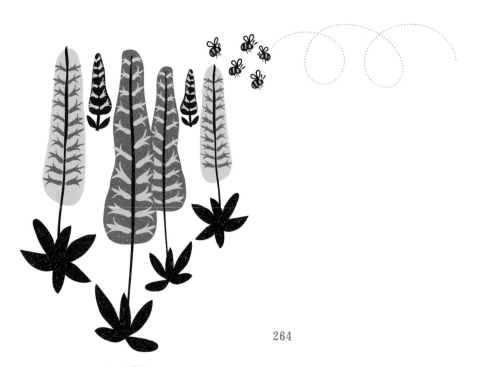

BEESWAX USES

TIP 458: *Use pure beeswax to make high-quality candles*

Making high-quality beeswax candles is an art form that is pleasing in several ways. A beeswax candle, correctly made, is nearly dripless. The odour is delicate and natural and the yellow flame accentuates the yellow candle. Entire books are available for the craft of candle making. Candle tapers can be dipped, poured or moulded. Candles from cappings wax make the finest candles, but any nicely made beeswax candles are highly marketable. Candle making is an old, established craft. Beeswax is flammable and the molten wax can cause serious burns and fires. Educate yourself and watch others who have experience before you give candle making a go.

TIP 459: *Select the correct wicking for the candle diameter*

The wicking used to feed the candle flame needs tested before turning out numerous candles. Using a chart to estimate the wick size in relation to the candle diameter is only an estimate. Even commercial candle producers use test burns to determine the proper proportion of wick to candle. The wicking used in votive candles will obviously be different from the diameter of the wicking used in tapers. A pickled wick burns much cleaner and brighter if not pickled. If the wicking is too large, the candle will burn very fast and result in large drips and misshapen candles. Alternatively, if the wicking is too small for the diameter of the candle, the candle will burn too slowly and the flame will not be able to consume the wax, eventually drowning in its own pool of wax. Ultimately, the candle flame will go out. Making beeswax candles is an enjoyable aspect of beekeeping, but correct wicking takes a bit of experimentation.

TIP 460: *Beeswax is great for lubricating wood screws*

✿ This natural lubricant truly makes a difference to lubricating the tip threads of wood screws. Even today, when screws are driven with drill-drivers, screws drive more easily – especially in hardwoods like oak or ash – and break less often. As the screw is driven, friction heat melts the wax and coats the screw and wood surfaces. Drawer sliders work more smoothly after rubbing a dry cake of wax on both the bearing surfaces of a drawer. Try also rubbing it on tool handles for polishing and improving grip.

TIP 461: *... And for lubricating a trailer hitch ball*

✿ The uses for beeswax are innumerable. As a dry lubricant, even more novel uses evolve. One such use is applying a dollop of beeswax to lubricate a trailer hitch ball. A few chips (about 2cm / ¾ in cubed) of beeswax can be put on top of a rusted connector ball and melted with a heat gun. As the wax runs down, wipe the ball with a paper cloth to evenly distribute the lubricant layer. Once thoroughly coated, the wax will reduce future rusting and will improve electrical connection surfaces.

TIP 462: *... Also for lubricating a pocketknife spring*

✿ A small chip of beeswax can be put into the spring mechanism to lubricate a pocketknife. The chip is melted with a heat gun and before solidifying; the knife blade mechanism should be flexed open and closed a couple of times. This will incorporate beeswax wax into the spring mechanism, making the blade easier to open. No lint or dust will stick to the hinge and any wax that gets on the knife can be nicely polished. Additionally, blades can be rubbed with beeswax to help prevent rust and make cleaning easier.

PROPOLIS USES

TIP 463: *Give propolis its due credit*

Beekeepers' awareness of propolis (bee glue) is as old as the beekeeping craft itself. Bees collect the basic materials from tree buds, sap flows or other botanical sources. Apparently, propolis acts as the bees' caulking filler compound and varnish sealer. It is biologically active; therefore, it suppresses bacterial growth. Its hive uses are numerous and no doubt more uses will be discovered. Propolis plays a major role in the concept of bee space. Anything less than 6mm (¼ in) is filled with propolis, while anything greater than 1cm (⅜ in) is filled with beeswax. Though the colour varies in a range of earth tones, the most common colour is dark brown. At temperatures above 20°C (68°F), propolis is sticky. At lower temperatures, bee glue becomes hard and brittle. Major uses of propolis inside the hive are to:

- reduce bacterial growth and seed germination,
- make wood surfaces slippery by coating with this natural varnish,
- reinforce combs and reduce colony vibrations,
- mask entrance odour and eliminate other entrances to the colony and
- entomb objects too large to move from the hive.

A healthy bee colony is one with its hive abundantly covered with propolis.

TIP 464: *Be careful when saying that bees 'collect' propolis*

Practically every beekeeping reference source will state that, 'Bees collect propolis from tree buds and other botanical sources'. In reality, bees either collect or produce the components – plant resins and gums, salivary enzymes, wax and foreign materials (mainly biological contaminants) – and mix the components within the hive to make propolis. Propolis as a product only exists within the hive and is a product specifically made by the bees. The gums and resins are difficult to harvest and store. There are few references to propolis being stored for later use, but propolis foragers will readily recycle propolis from old equipment. Within the warm hive, the product will soften and can be recycled.

TIP 465: *Evaluate human propolis health claims closely*

The advertised uses of propolis are too numerous and diversified to systematically list here. Taken at face value, it appears that propolis is beneficial to nearly any human malady. Some of the more common posted uses are for: skin and digestion tract inflammation, viral and bacterial diseases, an emollient for skin burns, sore throat lozenges and as an allergy treatment. Many other reported uses have not been mentioned. Clearly, propolis potentially has therapeutic effects on a wide range of issues, but scientific evaluation and subsequent proof has been elusive. The composition of propolis varies through the season, from region to region and from hive to hive.

TIP 466: *See if you like propolis toothpastes*

Propolis toothpaste is one of the more popular products that incorporate this little-known hive product. Sources for propolis toothpaste are easily found on the Internet. Some of the comments are: Propolis toothpaste can be used to clean teeth, freshen breath and protect from inflammation, gingivitis, gum disease, cavities and canker sores. Propolis mouthwash products are also available. The price per tube can be significant, ranging from £5–15 per tube. Medical endorsements abound and there is rational expectation that propolis can be beneficial in oral hygiene regimes, but, once again, the consumer will have to be the final judge.

TIP 467: *Leave chunks of scraped propolis for foragers*

Honeybee foragers will diligently work with old, sun-warmed propolis and will take small loads of repurposed propolis back to the hive for normal propolis uses. In the spring, large soft globs of new propolis can sometimes be found at the hive entrance and around extraneous openings. Typically, in order to keep the hive components and entrances clean and clear, beekeepers will scrape this product away. Leaving some of these clumps of semi-soft propolis in the sun, maybe on top of one of the hives, will make it easier for propolis foragers to find and will be interesting for you to watch.

TIP 468: *It's tricky to remove all propolis from your hives*

🐝 In the past, propolis was considered to be a hive nuisance. This bee glue product has tormented beekeepers from the outset. It made hundreds of hive designs impractical and even now it gums up hives that are not frequently managed. So why would the recommendation not be to scrape all this bothersome gum from hive equipment? The answer is because a healthy productive hive wants it there. There are good reasons for this hive product that are known only to bees. Like drones, propolis is a fundamental aspect of a prosperous colony. If beekeepers go to great efforts to remove all propolis, the colony will expend significant energy to gather new gum and constitute new propolis.

TIP 469: *Use plastic propolis traps to gather a larger crop*

🐝 A vigorous colony seems to be obsessed with filling the small cracks and crevices that provide hiding spaces for early stages of wax moths and small hive beetles. During spring and early summer and a bit during autumn, forager bees will gather the gum needed for mixing a batch of propolis and will seal cracks, crevices and other hive junk, as well as reduce the colony entrance. During those times, the beekeeper can put on propolis traps and harvest a meaningful crop of this lesser-known hive product.

Plastic grids, looking a lot like a queen excluder and containing hundreds of openings too small for bees to squeeze through, are put on the top of the colony just beneath the inner cover. The bees seem to hate those small open spaces and will fill them with colourful, new, nice-smelling propolis. The grid is removed and frozen overnight. After freezing, the grid (brittle with frozen propolis) is held over a large shallow pan and twisted. The clean brittle propolis will rain into the pan. Or, a shallow board of nearly any size that will fit on top of the colony can be cut with 3mm (⅛in) saw thickness.

TIP 470: Try a heat gun for softening propolis in hive bodies

A heat gun can be used to soften propolis stuck in hive equipment. Once softened, it is much easier to remove from the equipment. True, it can be scraped while hard and brittle, but the stuff will shatter and fly all about. The heat gun makes for less work and the room will have a pleasant smell. Without bees on them, stuck frames can be heated and removed from old equipment without too much breakage.

TIP 471: Expect propolis markets to be specialised

Simply stated, propolis markets are elusive, but widespread. Apparently there is not a specific commercial pathway for this product. Some bee operations have invested time and expertise into developing systems that will powder propolis so it can be made into capsules. In many instances, individuals who consume propolis for personal health reasons will buy raw propolis specifically from beekeepers. Propolis's uses are diverse and eclectic and are usually a minor component. Clearly, markets for quality propolis (not hive scrapings, which contain too much wax) are available, but the beekeeper will have to search for them with offers of product samples and an assurance of quantity. Marketing propolis is not like marketing honey, pollen or wax.

TIP 472: Protect floors and countertops from spillage

Propolis is tough stuff to get off clothes, the floor and equipment. Common alcohol will cut warm propolis a bit, but in many cases the stains, probably caused by tannins, will remain. Accumulations of propolis and wax on floors in honey-processing rooms are legendary. Putting down temporary floor protectors such as plywood sheets is a good idea. Once propolis is on the floor and ingrained, it is easier and more efficient to scrape off with a repurposed paint scraper on a long handle rather than attempting to chemically remove it with alcohol or strong cleaning agents.

TIP 473: Know that nurse bees use propolis throughout the brood-producing season

Little is known about specifics, but apparently nurse bees use small amounts of propolis to clean and polish a brood cell in preparation for the queen to deposit an egg. It has even been reported that the queen will not lay eggs in unpropolised cells. The apparent reason is that the biological activity of propolis offers antibiotic protection to the developing larva. How and where propolis supplies are stockpiled in the colony is unknown. It seems that, over time, the entire hive insides – all surfaces – are covered in this protective, multipurpose natural product. Expect more scientific advances to be made on this hive product in the future.

TIP 474: *Save some propolis for teaching purposes*

When fresh, propolis is pliable. Within a few weeks, it becomes hard and brittle, but it will always retain a unique, but pleasant odour. Audiences, both young and old, who are not familiar with bees may never have heard of this hive product. Introducing them to propolis is always fun. Since the dark brown ball of propolis has a questionable look about it, passing a hardened blob around the room and suggesting that each person sniff it will raise the eyebrows of the fainthearted. It is an educational moment in beekeeping.

TIP 475: *Expect hives to get slightly damaged in moves*

Propolis as a meaningful hive glue is deceptive. The hive may seem to be soundly stuck together, but propolis's resistance to being drawn, stretched or formed after curing means that the naturally brittle glue tends to abruptly break or crack if subjected to stressing forces. In hive use, propolis is not actually glue but rather an excellent sealant that simply feels and acts like glue. The disastrous effects of treating it like a glue are universal and often comical. When beekeepers – it usually takes two to move a colony – feel that the colony is 'glued' well enough for the colony to be moved without any straps, without warning the colony will unexpectedly break into its component pieces. This event will not be lost on the bees inside the hive. Always be aware that hives not strapped together for a relocation event are likely to break apart.

POLLEN AND YOU

TIP 476: *Help others understand that it is 'bee-collected' pollen*

New beekeepers or members of the public often misunderstand the terminology 'bee pollen'. Bee pollen is a pollen that bees collect from plants and blossoms. In no way and at no time do bees actually produce pollen (see chapter 3). Rather, the bees are always gathering pollen and storing it near their brood nest.

TIP 477: *Realise pollen nutrients differ from region to region*

It is true that pollen contains essential amino acids and many vitamins and minerals that promote general wellness and good health programmes. The analyzed nutritional values can vary from region to region and, indeed, some pollen producers in specific areas promote their pollen crops as being particularly healthy. In some advertisements, points are made that pollen is an environmental indicator of sorts. It has been reported that air or ground pollutants or environmental toxins show up in the pollen crop; ergo, eating the pollen crop will transfer these toxins to the consumer. While this may be generally true, no doubt other local vegetable and fruit crops will be just as affected. This is not a peculiarity specific to pollen consumption. Pollen nutrients can be found in other more common foods, but even so, many people feel that pollen is a particularly good food source. Ultimately, it is an individual call, but the consumer should expect conflicting claims and some nutritional confusion in regards to pollen consumption by humans.

TIP 478: *Do try eating local honey for allergy control*

❀ Consuming local honey and pollen to alleviate allergy conditions is a common recommendation within the health food community. Certainly, it may help in some instances, but in general, pollen collected by insects such as bees is denser and heavier than pollen from wind-pollinated plants. Many people react to windblown pollen from ragweed or grasses and are not particularly affected by pollen from sources such as apples or melons that are insect-pollinated. Whether or not to try local honey and pollens to help with personal allergies is a personal decision.

TIP 479: *Review the claims for specific pollen uses*

❀ Some athletes take bee pollen capsules to improve their stamina and ability. In a different instance, pollen has been reported to be helpful in controlling chronic prostatitis. As with these and other medical claims, monitor the sources closely. Though the recommendations may be common, the scientific evidence is not always strong.

TIP 480: *It's best to consult a doctor before eating pollen*

❀ No doubt many people routinely eat pollen on a regular basis and many of them never have any problems. Bee pollen is felt, in some circles, to be a premium health food, but there are medical concerns for some consumers. Eating bee pollen can cause an allergic reaction even in individuals who are not allergic to bee stings. Common symptoms that may indicate a food-induced allergic reaction include skin flushing, wheezing, itchy throat, hives, coughing, vomiting, diarrhoea and/or headache. In the worst cases, a severe allergic reaction may result that could include anaphylactic shock. If you develop difficulty breathing or swelling of the face, lips or tongue, seek immediate emergency medical attention.

TIP 481: Start off by eating small amounts of pollen

If you do decide to consume bee pollen, the general recommendation is to begin at a low level and increase over time. A common recommendation found on the Internet is to put a single pollen pellet beneath the tongue and allow it to dissolve. If no allergy issues arise, increase the number of pollen kernels over the next few days. A typical maximum level is a tablespoon, though some enthusiasts routinely take two or more tablespoons per day. The health benefits are widely distributed, but actual clinical tests are sometimes lacking. Begin with small amounts and increase over time. When determining how much pollen to consume at one time, it is best to err on the cautious side.

TIP 482: Perhaps expect occasional digestive upsets

For an unfortunate subgroup of the population, intestinal issues are too common. In some consumers, eating bee pollen may cause an upset stomach or vomiting. Bee pollen in susceptible people may also cause additional gastrointestinal problems such as diarrhoea, stomach cramps, and nausea. It appears that most people are perfectly fine eating reasonable amounts of natural bee pollen, but for a few there are true concerns. That group, if they have tried pollen, would already know that they have intestinal issues and take proper precautions. Incredibly, there are abundant testimonials and procedures about taking bee pollen to address their same intestinal disorders. The unanswered question is, 'Does bee pollen cause digestive disruptions or does bee pollen help with digestive disruptions?'

TIP 483: *Be wise to overstated nutritional and health benefits*

Consuming bee pollen as a curative and as a nutritional food is a matter of personal choice. There are avid supporters of bee pollen as a therapeutic food who believe strongly that bee pollen is a super-food. However, in most instances, typical scientific results either have not supported or only weakly supported some of the claims for pollen's health benefits. In some instances, the same data can be used to draw different conclusions. Be sceptical if the claims are broad and diverse for pollen's benefits. Although it is clear that pollen is a beneficial natural medicine in some instances, it is not always clear exactly what those instances are and how much should be consumed.

TIP 484: *Consider pollen a 'good' food, not a 'perfect' food*

To many, pollen is a near-perfect food and it does have supportable nutritional values. The protein, fat, vitamin and mineral content of pollen is comparable to beef or chicken and is similar to fruit and vegetables like cabbage, tomatoes, baked beans and apples. In bee pollen, vitamin levels for thiamin and riboflavin are actually 10 times higher than the levels found in the common foods listed above. This characteristic is interesting – especially for vitamin-poor parts of the world – but too much pollen would have to be consumed to meet daily requirements. However, other studies found that while the food values were present, natural pollen was essentially indigestible by the human system. This is a frustrating conclusion. Bee pollen's purported benefits are well known but poorly scientifically documented. Continued research is needed to determine the true benefits of pollen.

TIP 485: *Be conscientious if collecting pollen for consumption*

Bees collect the pollen product from outdoor plants, and transport it back to the hive where the beekeeper traps it. Pollen does not have the storage stamina that honey has. By design, honey is a stable product with hygienic systems in place. Pollen is much more temporary. Bees store enough for their dietary needs and not much more. Stored pollen loses its potency. Collect pollen from the safest sources available and away from pesticide application areas. Once collected, clean the pollen. This is usually done by hand, though compressed air may be used to blow detritus from the heavier pollen pellets. Larger quantities of pollen should be solidly frozen – usually in small quantities. Freezing prevents nutritional degradation and kills all stages of the wax moth. Pollen is usually sold in glass jars that are kept from the light. Generally, pollen is not washed but sold almost as the bees collected it. Keep it as clean as possible.

TIP 486: *Be prepared for powdery, clingy-tasting pellets*

Honey is known for its sweetness and individual pollen grains are made sticky with nectar so the bees can pelletise them. It would seem logical that raw bee pollen pellets should be sweet, but in fact, they are not. To most novice eaters, raw pollen pellets taste powdery and clingy. Dried pellets are crunchier. Over time, a taste for the product can be developed and consumption becomes more pleasant.

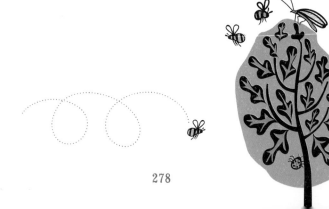

TIP 487: *If so inclined, try bee bread*

🐝 Bee bread is a coveted food in parts of the world, but is rarely seen as human food in many countries. Bee bread (stored pollen mixed with salivary juices and either nectar or honey) is partially fermented by the storing process, making it more storable for longer periods of time. The fermentation process also makes more of the nutrients available to the consumer. This same partially digested attribute also makes stored pollen more nutritious for honeybees. When traditional honey gatherers are raiding a natural bee nest, bee bread is commonly eaten on the spot. However, combs of the bee product can also be taken back to the dwelling, where it will keep for days, or can be sold as a local delicacy. It can also be eaten with a bit of bee brood or can be soaked in honey and eaten as a sweet.

TIP 488: *Note that pollen can be good for animals too*

🐝 Many of the characteristics described above for humans are thought to be beneficial for animals also. For instance, some select horse diets have bee pollen additives. It appears that horses grazing on natural grasslands occasionally eat blossoms containing pollen. Some horse owners feel that adding bee pollen may help horses with allergies and runny eyes and provide a natural grassland substitute. Following the improved human sports stamina theory, some feel that performance horses have increased endurance after eating bee pollen. Also, pollen is thought to improve breeding vitality and reduce stress. Dog diets containing pollen are readily available. Comments on general dog health and coat lustre are frequently given as reasons for including it in a canine diet. Chickens and pet birds and pigs have been reported to profit from having pollen added to their diets. Apparently scientific studies neither confirm nor deny these dietary attributes, but individual experience and opinions are frequently firmly held. Just as with human consumption, try not to overdose your animals or livestock with too much when you are trying it out for the first time.

BEE VENOM USES

TIP 489: *Expect pure venom to be difficult to get*

Bee venom is, no doubt, the most difficult of all the bee products to accumulate or collect. Few beekeepers routinely perform the task. Venom-collection devices have been designed that produce a mild pulsating shock to returning bees as they land on a battery-charged electrical wire grid. The stunned bee involuntarily stings through a thin, stretchy, plastic film much like plastic wrap. The venom from the extruded stinger is squeezed on the bottom side of the plastic film as the bee withdraws her stinger during the brief interval when the electrical charge is not on. The bee is not killed by the process. Later the film is removed and taken to a lab, where the venom crystals are collected by scraping or washing them off. When the bees sting through the film, alarm pheromone (iso pentyl acetate) is released and individual bees that are being mildly shocked are highly agitated. Anyone planning to build such a device should expect the colony to become hyper-agitated. Clever beekeepers can get plans from the Internet; a few models are also available commercially.

TIP 490: *Collect bee venom beneath a ventilated hood*

Though most beekeepers become remarkably resistant to the effects of stings, if either the beekeeper or an immediate family member breathes the bee venom, serious anaphylactic shock can occur. The venom is normally collected beneath a ventilated hood or at the very least collected outside. Though beekeepers are particularly at risk, no one should breathe bee venom directly. After all, it is a poison. The primary purpose of the collected venom is to desensitise individuals who are allergic to honeybee stings.

TIP 491: *Be aware of bee venom use and the claims*

🐝 Apitherapy, or the use of bee stings to address human aliments, is several thousand years old. References to bee sting therapy exist in ancient Egyptian and Greek writings. There are abundant anecdotal references to bee stings helping with various health issues, but there are few actual double-blind research results that support the use of bee venom therapy (BVT). Apparently, some beekeepers noticed that their arthritis improved after taking a few stings during routine hive manipulations, and from that awareness the concept of using bee stings for arthritis took hold. The difference between bee venom use and interest in other hive products, such as propolis or pollen as a medicine, is that overseas at least, there's more professional interest in bee venom. Possibly one of the reasons is the abundance of so many success stories and the testimonies of several trained medical professionals there. Interestingly, recently released details of a new study assert that bee venom can kill the human immunodeficiency virus (HIV). Researchers at the Washington University School of Medicine in US claim to have demonstrated that melittin, a toxin found in bee venom, destroys HIV.

TIP 492: *Explore bee venom use in cosmetics and skin creams*

🐝 The toxin melittin in bee venom is the active ingredient in some very expensive cosmetics that are thought to mimic the action of a mild bee sting. It is thought that the cream causes the body to direct blood towards the 'stung' area and stimulates the production of collagen and elastin. Collagen strengthens skin, whereas elastin keeps the skin tight and causes it to spring back into shape after being pressed. Bee venom absorbs through the skin and has been described as an alternative to Botox injections.

ROYAL JELLY USES

TIP 493: *Experiment with advertised royal jelly uses*

❀ Royal jelly is a white, thin, chalky-looking secretion produced by the hypopharyngeal gland located in the head of honeybees. Each brood cell containing a larva is given a ration for three days, after which workers and drones are put on a different and slightly reduced diet. The queen, however, is copiously fed this diet continuously for her entire 16-day development period – hence the name 'royal jelly'. Alternatively, workers and drones are fed at rates that result in the jelly being consumed immediately and without surplus. Both the amounts and the thickness may vary depending on geography, season and climate.

Royal jelly is mostly water (70%), but also contains a complex mix of sugars (10–16%), fats, (3–6%), proteins (12–15%) and trace elements, minerals and vitamins (2–3%). Some reported uses include treating stomach ulcers, liver disease, asthma, skin disorders, stomach ulcers, premenstrual syndrome, menopausal issues and kidney disease. Royal jelly is also thought to lower cholesterol levels and is even promoted for reducing the effects of aging and helping the immune system.

TIP 494: *... Though, don't expect to generate meaningful income producing it*

❀ This beehive product is presently extremely labour-intensive. There is no other way to produce it. It is commonly produced in areas of the world where the income from this speciality product can be produced by hand labour. Production on any scale is impractical for most beekeepers.

TIP 495: *Know how to produce queen cells*

🐝 By far the largest amount of royal jelly – about a drop per cell – is produced from queen cells. Developing queens are given large amounts of the royal jelly, while workers and drones are given smaller amounts of this nutritious food. Very large numbers of queen cells must be produced to produce meaningful amounts of jelly. On the fourth day of the 16-day development period, the jelly is removed from the queen cell with a small spatula and is immediately frozen. The larva in the queen cell does not complete her development. A healthy productive colony can produce about 0.45kg (1lb) of royal jelly during the 5–6 month season.

TIP 496: *Research bee product as human medicines*

🐝 Nearly all the tips that relate to using honeybee products as human medicines have been given in cautious terms. For many people, hive products such as propolis, pollen, royal jelly and bee venom are components of holistic or alternative medicines that are seen as natural and beneficial. The claims are so numerous and diverse that the scientific community does not have the resources to analytically check each claim for every product. No doubt some claims are valid while others are not. Even more confusing is that some products may work for some people but not for others. In specific cases, such as venom being used to desensitise allergic individuals, the uses are clear and beneficial. Individuals will have to decide if the product claims are worthy of review and subsequent use. Always inform your doctor if you want to try such treatments.

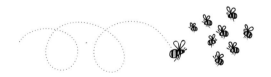

BEE BROOD USES

TIP 497: *Explore bee brood as edible and nutritious*

Bee brood is the least known and least revered beehive product. Yet next to honey, bee brood is probably the second oldest in importance to early honey hunters. Since ancient times, people in Africa and Asia have been ravenous honey eaters. Long honey-hunting trips were taken to help meet the nutritional demands of the group. While honeycombs were taken back to the village area, brood and bee bread were eaten immediately. Even today, bee brood and bee bread is sold in local markets in parts of Asia.

TIP 498: *Note that brood is an uncommon food source*

In the developed world, the intentional consumption of insects for food (entomophagy) is uncommon. A common estimation is that the average US citizen unintentionally eats as much as 0.9kg (2lb) of insects and insect body parts per year. It is simply impossible to produce commercial crops of rice, corn or wheat without the occasional insect part getting past detection. Is the consumption of the occasional insect body part a health issue? No, not really. In fact, if the same person harvested their own crops from their personal gardens, they would expect an occasional insect or spider to get harvested, too. If people were more accepting of insects as food, pesticide rates could be lowered and production costs could be reduced; but few authorities are expecting the public to readily and easily begin to see insects as food sources. Consequently, bee brood or even bee bread is not likely to become a significant beehive product anytime soon.

TIP 499: *Eat larvae and pupae, but not adult worker bees!*

Nowhere in the world are adult bees generally eaten as a routine food source. Since adult bees are necessary to either build or rebuild colonies and procreate the species, adult bees are considered valuable even if they are not kept in artificial domiciles. Secondly, the brittle exoskeleton, made mostly of crunchy chitin, makes the adult insects difficult to chew and nearly indigestible. Additionally, the venom can result in a burning sensation and other glands also contribute an unpleasant taste.

Normally, the larvae or pupae stages are eaten either with the wax comb or taken from the brood combs. The larvae and pupae are frequently deep–fried and impart the flavour of walnuts. In Nepal, bee brood is crushed in a coarse bag and the juice is gently stirred and heated. The results have a texture like that of scrambled eggs. Patties can have either boiled or steamed larvae crushed and added to the blend. Though initially repugnant, much more consideration should be given to bee larvae and other insects as food. The production of insect 'micro-livestock' is much less demanding on the environment and the protein value is as high as any present common food source.

TIP 500: *Put discarded bee brood to good use*

If larvae or pupae have to be killed, adult bees will readily recycle those dead bees as a food source. Nearly all of the larvae will be consumed, but some of the harder parts of the pupae will have to be discarded. Birds, especially chickens, are fond of bee brood. Wild songbirds are eager consumers of bee brood. There is little chitin and the immature bees are soft and palatable. Beneficial insect diets frequently have honeybee brood as an ingredient. The artificial diets used to attract beneficial insects such as ladybird beetles, lacewings and ladybird beetles include honeybee brood.

INDEX

USEFUL WEBSITES

APISERVICES
www.beekeeping.com

BRITISH BEEKEEPING
ASSOCIATION
www.bbka.org.uk

DEPARTMENT FOR
ENVIRONMENT, FOOD AND
RURAL AFFAIRS (DEFRA)
www.defra.gov.uk

FOOD STANDARDS AGENCY
www.food.gov.uk

INTERNATIONAL BEE
RESEARCH ASSOCIATION
www.ibra.org.uk

JAMES E. TEW'S BLOG
www.onetew.com

KEW GARDENS (POLLINATION)
www.kew.org

LONDON NATIONAL HONEY
SHOW
www.honeyshow.co.uk

NATIONAL ARCHIVES
– UK LEGISLATION
www.legislation.gov.uk

NATIONAL BEE UNIT (NBU)
/BEE BASE
www.nationalbeeunit.com